03/18

D1487873

DISCARDED

DEAR FAHRENHEIT 451

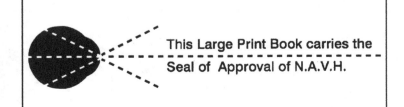

This Large Print Book carries the
Seal of Approval of N.A.V.H.

Dear Fahrenheit 451

LOVE AND HEARTBREAK IN THE STACKS: A LIBRARIAN'S LOVE LETTERS AND BREAKUP NOTES TO THE BOOKS IN HER LIFE

Annie Spence

THORNDIKE PRESS
A part of Gale, a Cengage Company

GALE
A Cengage Company

Farmington Hills, Mich • San Francisco • New York • Waterville, Maine
Meriden, Conn • Mason, Ohio • Chicago

Copyright © 2017 by Annie Spence.
Thorndike Press, a part of Gale, a Cengage Company.

ALL RIGHTS RESERVED
Thorndike Press® Large Print Peer Picks.
The text of this Large Print edition is unabridged.
Other aspects of the book may vary from the original edition.
Set in 16 pt. Plantin.

**LIBRARY OF CONGRESS CIP DATA ON FILE.
CATALOGUING IN PUBLICATION FOR THIS BOOK
IS AVAILABLE FROM THE LIBRARY OF CONGRESS**

ISBN-13: 978-1-4328-4839-2 (hardcover)
ISBN-10: 1-4328-4839-9 (hardcover)

Published in 2018 by arrangement with Macmillan Publishing Group,
LLC/Flatiron Books

Printed in Mexico
1 2 3 4 5 6 7 22 21 20 19 18

This book is dedicated to the other six Authiers from the double-wide that was always taking in more people and animals than there was space for, who raised me and had such white-hot pride for me I rarely suffered the cold of my own doubt.

CONTENTS

7

9

I absolutely demand of you and everyone I know that they be widely read in every damn field there is; in every religion and every art form and *don't* tell me you haven't got time! There's plenty of time. You need all of these cross-references. You never know when your head is going to use this fuel, this food for its purposes.

— Ray Bradbury

INTRODUCTION

Dear Reader,

Welcome to *Dear Fahrenheit 451*. Shall we beg — wait, I know you guys! Do you remember me? I'm your public librarian! I walked you over to the Murakami that time. I helped you get the DVD about exploring New Zealand and you came back and told me about how wonderful your trip was and we both got tears in our eyes. Remember when you said you paid my salary and mumbled "bitch" under your breath when I wouldn't do your kid's research paper for them? I'm *that* bitch!

I know all of you — because librarians love getting to know their communities: from *Junie B. Jones* Kid to Conspiracy Theory Andy! If I hold up my magic mirror, *Romper Room* style, I can see each and every one of you reading this right now! I see Geoff, who always says he's picking up his Regency romances for his sister (no judgment, Geoff!), and I see Donna, who reads philosophical horror novels as fast as I can supply them. I see Carol, whose grandson bought her a tablet and then apparently went into the witness protection program before he could help her figure out how to use it. (In fact, I see *all* the doting millennials who pat themselves on the back for giving expensive devices to their elderly relatives and then go back to college without explaining how to download an e-book.)

But as close as my connection is to all of you, your literary preferences and Internet habits, there is a population I know even more intimately: the stacks. Librarians aren't just reading while we're sitting at the reference desk. We curate the collection by providing a fine balance of items patrons need to be well-rounded (poetry, *Consumer Reports*) and items they request that we buy (more seniors' yoga on VHS). We also decide when a book is no longer needed and has to be "released" (two points if you got that *The Giver* reference). Professionally, we call this process "weeding" the collection. Personally, I call it "book breakups."

I know books on a deep level. So deep that, over the years, I've found myself talking to the books. Only in my head, because

I'm not crazy; but, inside my head, I talk to them in letter form, because books are fancy and need to be formally addressed. It used to be just at the library, while I was weeding or when I would come across an old friend — I mean, book. But now I seem to do it every time I look at a bookshelf: at my mom's, at a dinner party, at the bar or on date night. Basically, if you've spoken to me in the presence of a bookshelf in the past decade, I wasn't paying attention.

And why shouldn't I talk to books? I've got a lot to say to them.

Reading has shaped me, guided me, reflected me, and helped me understand and connect with, and this is not hyperbole, HUMANITY. If you picked up this book, it's because, somewhere in the past (and more in the future, if I have anything to do with it) a book has changed your life. Well, mine too, dear reader, mine too. I grew up in a small rural Michigan town. I was the youngest of a big family, living in a tiny house that was overflowing with people, stray dogs, love, and saltine crackers. We didn't have a lot back then, but we did have the library, and its books showed me a bigger world. I know that sounds confusing, because you're, like, wait, bigger world? Aren't you just still hanging out at the library? Did you ever even leave? I only mean that books have shown me some amazing things. They've thrilled me and soothed me. They've told me when it was time to give up on them. They've helped me not give up on myself. Reader, for all of the silliness and good goddamn fun involved in writing a book that talks to books, I know you'll believe me when I say that these books have talked right back to me.

And if this book you're holding could talk? It would say that it wants you to connect to it, to laugh with it, and to walk away with a whole new list of other books that you can't wait to get involved with. Happy reading.

Your Ever Lovin' Librarian,
Annie

Dear Librarians,
Please don't weed me.

Love,
Annie

■ ■ ■ ■

I.
BOOKS —
THE LETTERS

■ ■ ■ ■

Good friends, good books, and a sleepy conscience: this is the ideal life.

— Mark Twain

Rule number one: Don't fuck with librarians.

— Neil Gaiman,
Gaiman's online Journal, 2004

FICTION--Tartt, Donna
--Growing Apart

Dear *The Goldfinch,*

We've grown apart. Or, I guess, *you've* grown apart. Like, physically. Your spine is torn to crap. The hardest part about this? I'm the one who did it to you. I love you so much, *Goldfinch.* Your language, your emotion, your suspense. Needless to say, the author picture on your back cover is the main reason I started parting my hair down the middle.

So I recommended you to everyone. I broke the Librarian's Reader's Advisory Code, which is to base your reading suggestions for a patron on their previous preferences, not my own. I broke it for you, Finchy. I recommended you to folks checking out Sylvia Browne dead-people-talking books and patrons asking where the Amish fiction was shelved and people who told me the last book they enjoyed was *Hatchet* by Gary Paulsen, which is, sadly, every third adult male who comes into the library. I'm not saying you won the Pulitzer because of me, but you may want to think about adding one more name in the acknowledgments when the next edition comes out. You feel me?

Unfortunately, your hard exterior couldn't protect you from the reality of the world outside these shelves. It was bound to happen. You're nearly eight hundred pages. And about a gazillion people cracked you open. Eventually, you cracked too. It's my fault. I shouldn't have sent you home with people who are used to reading mass-market paperbacks. That's something I have to live with.

I know you are a book that only feels fulfilled when being read and admired. You'd be too ashamed to sit next to your other copies as busted up as you are, and there's nothing book glue can do for you now. You don't smell or anything, if that's a consolation. I'm taking you home with me. You'll sit right next to your old pal *The Little Friend,* on a browser-friendly shelf above the record player where my friends will look at you with great reverence before declining to borrow you because they are too busy to read (I know, they're fools). I'm the only one who truly knows you well enough to notice how fragile you are on the inside. No one but you and I will ever see the duct tape holding you together or the DISCARD stamp on your title page. I promise you that.

Seriously Forever Yours,

Annie

FICTION--Tolstoy, Leo
--Classic Russian Literature
--*The Bachelor*
--Choices

Dear *Anna Karenina,*

I feel like I don't even know you. Maybe that's why I find it so difficult to say: I've been seeing someone else.

Geez, I'm sorry. I know I've led you on. I asked my friends about you. I checked you out more than once. You came home with me. You stayed for a month! But while you were on my coffee table, looking so earnest and so very long, *Eleanor & Park* by Rainbow Rowell was in my bed. And then some Megan Abbott mysteries. And then Dolly Parton's autobiography. Twice.

I tried, I really did. Once, I even picked you up and held you. I kept you on my lap while I watched *The Bachelor.* And you made me feel better. Like I wasn't just some faceless citizen of Bachelor Nation. *I read Russian literature!* I thought to myself. *I'm just smugly observing this show until the next commercial, when I will begin my scholarly analysis.* But then I kept watching through "After the Final Rose."

Anna, I don't have one unkind word to say about you — because I haven't read you. Perhaps, it's just not our time. There will come a day, probably, when I get a hankering for a bleak 864-page novel translated from Russian. But until that day, back to the shelves you go.

I tried to look up "goodbye" in Russian, but it's really hard to spell. So, just —

Goodbye,
Annie

FICTION--Eugenides, Jeffrey
--Creepy Stories
--Creepy Love for Creepy Sto-
ries

Dear *The Virgin Suicides,*

Congratulations on your fifteenth consecu-
tive year as my favorite book. To mark this
commemorative anniversary, I'm writing
you a love letter. It'll be moony goony
nonsense compared to your perfection, but
the thing you're perfect about is conveying
imperfect love. So even though this is going
to look a little bit like pen puke, I hope
you'll appreciate its sincerity.

Here goes.

I love that you have no plot and an electric
story at the same time. The five Lisbon
sisters commit suicide in the suburbs of
Detroit; their neighbor boys loved them and
couldn't understand them. We know that in

your first few pages. Nothing else happens. Except everything. Except tiny beautiful moments of arms barely touching, and records playing over the phone, and sad math teachers, and goal-line chalk striping a beautiful girl's back — all of the minutiae that composes lives and somehow adds up to death.

I love every one of your fucking golden sentences. They are slam-you-shut-and-clutch-you-against-my-chest sublime. The description of the adornments that spill around the teenage girls, and the entire swooning Trip Fontaine passage, oh *man*, I wanted him to save Lux Lisbon so bad. Then, that last paragraph made me want to collapse on a fainting couch and linger for the rest of the day with your delicate memory.

I love that *after* I read you, every time, my own everyday movements and the quotidian moments of my life feel more beautiful. That's the mark of a lovely book. You make me want to never look at my phone again — to abandon Facebook in favor of old astronomy books and nature guides. I just want to brush my hair languidly in front of the mirror, sift through old costume jewelry, hold hands, and listen to way more Bread.

It's more than that, though. I feel like you

get me. Like, *get* me. I don't feel like you were written *for* me. I feel like you were written FROM INSIDE OF my psyche. The hazy gaze with which you look back on suburban Detroit is the same lens I was spying through growing up. My folks having moved from Detroit to rural mid-Michigan before I was born, those cities and the people left behind were a dreamy, mysterious world that existed in the Before Time of my parents' lives.

I was enamored with and naive of "downstate," as we called it, in the same way your narrators were always reaching for and never quite grasping the Lisbon sisters. The "fuzzy aura" that Trip saw surrounding Lux echoed the one I saw glowing around my older cousin Melanie when she came to visit from Detroit. The music the boys play for the girls after Mrs. Lisbon makes Lux burn her records was the music my dad put on our record player, and when he listened he got a far-off look in his eyes.

When I left for college, a whopping forty-five minutes away from home, I found myself drawn to the city girls in my dorm hall. Girls who called their purses "bags" and had gone to foreign countries for senior trip. Who had been to concerts that did not take place at the fairgrounds. Who owned hair straighteners and manicure kits. Who took for granted all the music and art and stories that grew from the same place they were from. These girls often felt compelled to give me life advice as they smoked next to their open windows, because they could tell I admired them and because they found the podunk homecoming queen vibe I gave off endearing. Absolutely all of it was terrible advice. But I don't mind because eventually one of those smoky window conversations, where I let them pretend to be Carrie from *Sex and the City* and wax nonsensical, led me to a conversation about you, *V.S.* I'd never heard of you or your movie, further cementing my charming rural bumpkin reputation. Anyway, they let me borrow you. From your first paragraph I knew that, finally, I could identify that feeling of reminiscing for a place I had never known but felt connected to. It felt like fate.

How else could I explain this kismet I feel? Oh, I know — you'll get this because it's another Michigan reference: you feel like the first time I heard "Against the Wind" by Bob Seger. Like, I'm seven years old, with no true drifter days under my belt yet, but I'm still swaying in that station wagon going, "Goddammit. He's got nothing left to burn. He's just livin' to run and runnin' to live." That's how close I feel to you. Seger close.

This year, I moved to the suburb you take place in. (Not *because* of you — I'm not that weird.) And, I'm reminded of you now that it's fish-fly season, which is when your tale begins. The bugs are blanketing store windows and cracking under our bare feet on the cement steps we sit on in the evening. The neighbors and the gas lamps in each suburban yard are already swathed in nostalgia for me, a mix of my own history and the fictional one you created. Our stories are braided together.

This is all to say, God, I think you're groovy.

Happy Anniversary,
Annie

GENERAL MATHEMATICS--Rogers, James T.
--Calculators
--Old as Shit

Dear *The Calculating Book: Fun and Games with Your Pocket Calculator,*

We never go out anymore. To be more specific: you. You are REALLY not getting out much these days. It's not that recreational mathematics isn't a thing anymore. I guess it's just that — how do I say this? Remember how on your book cover you ask if we have ever wanted to greet a friend electronically? People have kind of figured out how to do that without turning their calculators upside down to spell "Hello."

We used to have so much fun together! Remember 80085 ("Boobs"!)? I couldn't get enough of that in the old days. But now when we meet, you say things like "A few extraordinarily rich men have displayed the quirk of never carrying any cash" and "Your calculator can give you advice on driving." You know about GPS, right? Have you even heard of Google? It's 379009 upside down.

This is hard for me. I still like your "electronic whiz kid" bravado. You've still got a bit of the ole charisma. But the kids think the picture of the calculator on your cover is, like, the first iPhone. I'm making a Brain Games display next month. There's a book from 1983 about the videogame craze you might like to meet. I'll put you two next to each other and hope that your "old-school" covers work some magic when the kids come in to play Minecraft. If that doesn't work, it's Free Box time.

L8R (That's LATER),
Annie

FICTION--Christie, Agatha
--Mystery Series
--Ladies, Old 'N' Sassy
--Sure Bets

Dear Miss Marple Series,

You guyssssss! I just want to thank you for
being there for me. Everybody loves you.
Seriously, everybody. I mean, people who
like mysteries — doy. But also, did you
know that truckers love you? You guys on
audio are like a gateway drug to reading for
truckers. Also, kids who read way above
their grade level and are bored with everything in the children's section. The cute 'n'
gawky ones who get big smiles when you
give them a book, and they sit down on the
floor and start reading it immediately. And
teens with helicopter parents who want to
make sure they aren't reading novels with
sex in them (as a rule, murder in a book is

a-okay with these folks). And millennials love you because they picture your main character as Mrs. Doubtfire.

You make my job so much easier on days when I spend the better part of an hour with a patron, placing stacks of books in front of them, shelling out the "If You Like" bookmarks like it's my job (which it is) and presenting each tome to a frowning face.

"Something historical. No, that's too historical."

"I definitely want death, but don't necessarily want to read about anyone dying."

"I'd like to be intrigued, but not confused."

And then, like a clammy grandmotherly hand gently smacking my cheek, your name comes to me. You are popular enough that everyone has heard of you, old enough that at least one of you is always available on the shelves. You're sassy enough for a chortle, and well-mannered enough that you won't offend church ladies or parents who monitor every library book their child checks out, apparently unaware of Snapchat. You can be read in any order. You're clever in an "oh yooooou" kind of way that doesn't make your readers feel dumb. And you're a "cozy" series that is also well written — the romping unicorn of the Mystery section.

You are my Back Pocket *Body in the Library.* My Anytime *At Bertram's Hotel.* My *Pocket Full of* Reading Pleasure. I got *The Thirteen Problems* and a disgruntled library patron ain't one. They'll always choose you. Even the dudes. Even the ladies that are mad at me because the Jodi Picoult book they want is still checked out and they are making me pick out other, not-as-good books for them as punishment.

Who would have thought a gossipy spinster from St. Mary Mead could bring us all together? Sometimes, I wonder if the library could get by on a collection that was just you, *A Child Called It, The Five Love Languages,* and some Rick Steves travel DVDs. Honestly, I think we'd make it at least a week before someone complained. But you're my favorite, Miss M. My good-time gal.

Props,
Annie

FICTION--Sparks, Nicholas
--Horses, Definitely
--Angels, Maybe

Dear *Dear John,*

You're on my shelves because a relative who shall remain nameless recommended you on three separate occasions, and I didn't want to seem judgy. Hey, turns out I'm judgy.

You've got horses on your original cover. That's cool. I came into my compulsory Girls Love Horses stage in adulthood, and I'm still riding it out — pun intended.

That's about all we have in common.

I read (most of) your prologue, and I just can't see this reader-book thing continuing between us. I'm pretty sure your main character got dumped while he was away in the army, so I just want to be up front with you and save you more heartache. You don't need any more heartache. For SURE.

It's not that I don't want to be romanced. But I need more. It sounds kind of obvious to say this because you're a book, but I want to be moved by your words. In the prologue you say, "Our story has three parts: a beginning, a middle, and an end." No shit, *John*. That's how that works. Give me something that I've never heard. Describe something I'm familiar with but never thought of as beautiful before. Or at least throw in some more equestrian scenes.

Anyway, not to beat an underutilized horse, but I'm donating you to my doctor's office. I don't know what I'm going to make up to say to my relative when she asks if I read you. I'm just guessing, given your author, that one character turns out to be an angel? I'm gonna hedge my bets and lead with that at the Christmas party.

So Long, Pardner,
Annie

FICTION--Niffenegger, Audrey
--Time Travel, in Literature
--Time Travel, for Real Though

Dear *The Time Traveler's Wife,*

I knew that I loved you, but I didn't remember the reasons. In public, when you came up in conversation, I spoke affectionately of you. But at home I ignored you. It had been too long since I'd read you, and we'd become more like acquaintances than reader and book. But then something you'd said once popped into my head, for some reason: "It's the reality that I want." So I picked you up, walked you over to the couch, and really looked at you again. I stared straight into your insides. And I fell in love all over.

You first piqued my interest in 2007 because your main character, Henry (sweet, busted-up Henry), was employed at the same library I was working for: the Newberry Library in Chicago. I was already geeked that one of my first library jobs was so prestigious: a 120-year-old research library with visiting scholars from across the world and six centuries of material in its collection is cool on its own; but when your *book job* is cited in a *book*? That shit is meta.

I loved reading you "on my break" at work and then visiting all of the spots you mention. I wandered around those same freezing-cold stacks and sometimes got lost, just like Henry. Well, not naked, though I thought about it.

Reading you brought out the romance and mystery of working at the Newberry: the velvet pads we placed under delicate books, the chaotic and genius scholars, the book rumored to be covered in human skin (touched it). For the first time, when I told people I was becoming a librarian, the response I got was "Oh, like that book!" instead of the too common "Hey little lady, your job is going to be obsolete. E-books. Home computers. Blah blah mansplain blah."

Also, I thought Henry and Clare were sexy. I was similar to twentysomething Henry when I was twenty-three, minus the involuntary time traveling, the opiates, and the punk music (which is everything interesting about Henry, I realize). And I identified with Clare's constant craving for her beau while she waited for him to come back from time traveling. Her longing looked to me like the pining that many of us romanticize when we're young, our idea of big love before we encounter it in the flesh, with all its pocks and scars.

Back then, I speed-read through your passages about Clare's art and the ways she sacrificed one passion in favor of her fervor for Henry. I think I thought something similar to "Who cares about your papier-mâché? You are independently wealthy with a hunky brilliant husband who time-travel visits you at different ages." If hashtags had been popular then, I would have just thought, "#richgirlproblems," but you really had to reflect in complete sentences in those days. Anyway, I was in the infatuation stage with you. I was in it for the stirrings and the promise of love.

But recently I read you again. This time I am married with a child, just as Henry and Clare find themselves when they have Alba. And you wrecked my shit. At twenty-three, you were a fantasy. At thirty-two, you're a mirror.

The time travel in real marriages is way less sexy. It's night meetings and laundry and aging parents, and running errands together counting as a date. It's binge-watching a show on Netflix when the other is gone and feeling guilty for not feeling guilty. And pining for your lover not when he's naked somewhere in the time continuum, maybe beating up your old high school boyfriend or whatever the hell teenage Henry was doing with/to himself in his room on pages 55–56, but pining for him when you've shaved your legs and put on decent underwear but then right as you're ready to get the bidness on, your kid has a night terror, or someone texts one of you a hilarious video, or you actually just want to finish your goddamn book.

And it's knowing that after surviving all of that shit and still liking each other, our "time travel" inevitably becomes cancer, car accidents, strokes, and senility if we're lucky enough to hold on to one another that long. There's more literary drama when Clare hopes that Henry doesn't time travel during their wedding ceremony or the birth of their child, but the emotion it evokes is the very sincere and honest worry of even the most mundane non-sci-fi marriage: Don't leave me. Please don't ever go away.

My second time through, I'm weeping for Henry each time he holds his daughter for maybe the last time. When he sees Clare before all the worry and stress of their life together wore her down, I cry for the ways a marriage and children can fill you to bursting and at the same time deplete you. And then when Henry sees Clare again when she's old, in her little painting shed with her sad but regal old-lady braid, I realize that Clare's art is what grounded her while Henry kept her up in the clouds, and you need both. You need *both*!

Then suddenly I'm shaking my husband, Michael, and saying, "I fucking love you. Don't ever travel to different years!" And he's like, "I'm just trying to watch this 2009 Blur live concert for the six hundredth time," which is the only nerdy little version of time travel he'll ever accomplish, and I love him even more for it, so I just sob on the floor until he asks if maybe getting some chicken shawarma would cheer me up. He really gets me, you know?

And so do you, *TTW*. From my heady young librarian-about-town days to now. I promise not to forget again. One day, when I'm an old lady with my own sad braid and a book shed that my son locks me in during the day because I've started confusing real life with novel plots, you'll travel back to me. I'll be journaling in my spare time, hallucinating that I'm Anne Frank or Bridget Jones, and my vigorous scribbling will shake the table enough that you'll fall from some swollen pile and the thump will wake me from my delusion. I'll remember you. And I can't wait to see what you'll mean to me then.

See You Soonish,
Annie

PHONOLOGY AND PHONETICS
`--Pronunciation, Proper`

Dear Pulitzer Prize–Winning Books,

Is it PULL-itzer or PEW-litzer? I never know. I hope it's not the latter. It's hard not to sound like a dick when you say it like that.

Don't Judge Me,
Annie

FICTION--James, E L
--No Means No
--Erotica, Sort of

Dear *Grey: Fifty Shades of Grey as Told by Christian,*

Whhhhyyyyy do people keep asking me if I've read you? Aren't you the same book as the last one of you I said I didn't want to read?

But nobody cares! They can't get enough of you. They read your first version. They'll read you. And in between they will have no desire to read anything else. It's like, without you, they would just rather be illiterate. It makes me want to shake readers and scream: YOU'RE SURROUNDED BY GREAT LITERATURE AND THIS SHIT ISN'T EVEN THAT DIRTY! I want to clothe Pablo Neruda poetry with your jacket and hand that out to patrons. And I don't care what you think, Greeeey. Your silk tie–anal bead sweet talk, or whatever it is you're saying to get on top of people's nightstands, isn't going to work on me.

Today yet another person asked me about you: "Have you heard of a book called *Grey*? It's like *Fifty Shades of Grey,* but from a Christian's perspective?"

So on top of having to politely smile when people say to me, "You're bookish, I've got one you'll love!," *now* I have to explain to a little old lady who only reads Karen Kingsbury novels what erotica is and watch her pretend to put it back and then pick it up again when I'm pretending not to look.

You made me say "erotica" to an old lady, *Grey!* I'm going to hate you forever for that.

I'm putting you out on the curb where you belong, and I hope someone drops you in the bubble bath they are sitting in when they read you.

You Nasty,
Annie

FICTION--Trigiani, Adriana
--Lovable Characters
--Needy Reader

Dear Big Stone Gap Series,

Where've you been all my life? In the hands of other readers is where, and I've helped put you there. I've sung your praises to patrons checking out Loretta Lynn music and Lorna Landvik books and other artists who don't even have double Ls in their names.

You're a series with a strong and lovable heroine named Ave Maria. You take place in Appalachia in the 1970s, so people who like historical fiction will enjoy your cultural details, but people who aren't that into history will still find you palatable. You're a good balance of funny and sad. All of this I knew from other librarians and patrons who enjoyed you, but I'd never read you myself.

Then last month, I saw someone looking at your first of four — *Big Stone Gap.* She was crouched on a stool next to your shelf. She had obviously been browsing and been so taken with you she couldn't stop reading. I knew because she had The Look on her face: the look people get when their brains are so engrossed that they don't care about their outward appearance. Because they're not out in public anymore, they're in whatever world they're reading about. It's beautiful. This is not to be confused with That Look, which is the look on people's faces when they're browsing the Internet and are *not in control* of their outward appearance. If you think I'm making this up, visit any space where there are more than five computers in a row. It will dismantle your faith in humankind. It looks like a tribute to Daniel Stern's facial expressions from *Home Alone.*

Anyway, I saw The Look and I wanted it. So I put *BSG* on hold, and when you came in last week, my own world stopped. I mean, whomp! I fell into your small mining town and didn't surface again until Jack and Baby Etta were sitting on the porch, gazing at the moon. You said that when we get what we want, time stops, and you were right.

I fell in love with Ave Maria: a book lover

and a daydreamer. Her character is more realistic than I had expected; the everyday loneliness she feels as the town "spinster" is palpable. And your other characters are just as lovable. Normally I resent oversexed librarians in literature — it's overdone — but I found Iva Lou's hunger for life endearing. And Theodore and Fleeta and Pearl — I wanted them to be my friends! Then Jack Mac, what a man.

I was surprised at the originality of your story. It's not just: a lady that's "gettin' up there" finds a man. Ave finds her secret self first, the woman she always wanted to be. She finds family and friendship. And she sleeps for seven days at one point, which, honestly, as the mother of a toddler, was more arousing to me than the romantic parts.

Now that I'm hooked, I've got to read the other three in the series. So I have to ask, are you in it for the long haul? Am I going to ruin this experience by going further?

When I pick up *Big Cherry Holler,* will Jack and Ave still be in love? I worry about those strong and silent men — they're not always silent because they're *thinking.* What if he's just dumb? What about Italy and Ave's father? Did Pearl go to college? Can Iva Lou's marriage last? Part of me wants to know, but part of me wants to preserve them the way they are, right there in Big Stone Gap.

I came over here to tell you guys, please don't let me down. We're in it together now.

Sally Forth,
Annie

FICTION--King, Stephen
--Are You Okay?, Annie

Dear *Misery,*

Jesus. Phew! We've been through the shit, eh?

Shit, man.

Shit.

Let me recover for a minute. I need a drink or a sedative or something. I'm a regular Paul Sheldon.

Okay, so until you there's only been one other book that I read wherein the whole time I was reading I kept my hand over my heart. That is a book of Robert F. Kennedy speeches. For you, I did the same. Totally different reasons, obviously. RFK filled me with hope and pride; with you, I was just clutching at my body to make sure my hand didn't get sucked into the page and hacked off.

Shit, man.

You are . . . disconcerting. Something ain't right with you. A writer held prisoner in the home of his number-one fan. I thought I could handle it. And, at points, when you were dishing out the wise writerly insights, I could. But, my God, what about Annie? How did you house a character as quietly, explosively crazy as Annie Wilkes without shriveling up? I can smell her. I can feel her ruddy cheek brush mine.

Not to mention that, just to freak my ass out more, you gave her my name. At first I was happy about this. Usually characters named Annie are dead but longed-for high school sweethearts or sassy red-haired foils for Kevin Costner in baseball movies. I was eager to see an Annie with a little more grit and girth. But hearing a character plead to *me* while he was getting his parts sawed off and cauterized was unnerving. I was like, I can't help you, dude. I'm just as scared as you. That bitch is crazy! Quit saying my name!!!

Worse than the amputations though, *Misery*? The rat squeezing. You know when you get so scared that for some reason your butt can't be touching your seat anymore? When your ass cheeks fold up into your thighs? That was me with the rat squeezing. And the licking. You know what I'm talking about. "Too much!" I screamed, to no one in particular. "Africaaaaaa!"

Then when that first trooper comes? Another buttclencher. I knew it. I knew when you said he was so young that Annie was gonna get him. But you couldn't just leave it at that, could you? I've got to think about him every time I mow the lawn now. Shit.

I'm sorry I threw you across the room at the end there. But you're too intense! I was turning your pages as fast and fearfully as Paul when he finds the oogy murder scrapbook. And I kept hearing my name. Then there were more rats. Then her fucking fingers were moving under the door! Get out, you dirty bird, she's coming!!!

You're a great book, but I gotta give you back to my sister. There's only room for one Annie in this house, and it's the one who's going to tuck her thumbs into her armpits every time she sees an electric knife for the rest of her life.

Shanks for the Memories,
A. Spence

BOOKS--Various
--Me, Well Excuuuuuse

Dear Fancy Bookshelf at a Party I Wasn't
Technically Invited To,

8:00 P.M.
Hey. I'm just going to stand here and look
at you for — forever, probably, because I
don't know anyone here except the friend I
tagged along with, and I don't want to cock-
block her. Are you familiar with the phrase
"cock-block"? I see you have a leather-
bound copy of *Pride and Prejudice* — it's
like Mr. Darcy.

I'm sort of a bookshelf flower anyway, so I
don't mind. You're so glamorous, with your
art books and color blocking and your
decorative sea-conch accessories. You've
probably got more to say than anyone else
here. For instance, from your *Pride* and

58

dainty shelf accoutrements, I can tell you lean more toward Austen than the Brontë sisters. I'm a Brontë girl myself. I like to brood. Next to bookshelves. Which brings us full circle.

Waaaiiit, this shit is catered?!?!

8:15 P.M.

Alright, alright, alright. Drinks 'n' snacks. Anyway, I'm just saying you're very well put together. My shelf at home is such a mess. Annie Dillard next to *Autobiography of Malcolm X* next to *The Shape of Me and Other Stuff,* which is not a weird sex manual, FYI. I just felt like I needed to make that clear.

My husband buys series in mass-market paperbacks. And our books don't really "go" together. If someone were to judge who lives in our house based on our bookshelf, they would probably guess we were socialist botanists, equally obsessed with the Beatles and the best clothes for our body types, a couple who collects classic novels from Dumpsters and owns a dog that occasionally gnaws the corners of our already-shabby collection. And that's simply not true. We have a cat.

Most of the books I truly love aren't even on my shelves, because I loan them out to other people or get them from the library and have to return them. I foolishly had Jonathan Safran Foer sign a copy of *Extremely Loud and Incredibly Close* to a boyfriend who wasn't able to attend the reading, and then when we broke up I let him keep it. Even my favorite authors — the pride of my collection — their copies are scruffy paperbacks with dog-eared pages. My Karen Joy Fowlers are so worn from use that you can no longer read their titles from their spines. My Faulkners look pretty good, though.

But you. You're an inspiration. You've really thought yourself out. Your pink shelf, especially. *Valley of the Dolls,* a Marie Antoinette bio, and *Chanel and Her World.* Beauty and controversy — I think that's what you're saying there. Is that what you're saying? Or is it like in college when you went to a creative writing workshop and everybody's going on about how postmodern a story is, but it turns out the printer just collated the pages out of order?

9:00 P.M.

Can I set this gin fizz on you? I'm going to put it right next to *The Great Gatsby* so I'll remember where it is. Drinking joke!

But seriously, I can't go overboard with

the alcohol because I tend to pontificate about reading and the social significance of the public library when I get drunk. Two drinks: funny work stories. Poop in the dropbox, Lady with the Face, the guy whom we caught looking at porn and eating a big can of sardines and we didn't know what to be more offended by . . . that kind of thing. But if I morph into telling *inspiring* patron stories, look out. I can give a rousing/annoying lecture on the benefits of getting your library card. I've shouted, "I disseminate information to the masses!" while being helped into a cab before. It's not pretty. Oh, do you have *Let Me Tell You, It Wasn't Pretty* by Diane Keaton? I don't see it. I'm going to go tell the host she should get it. That would look really good on you.

Where's my drink?

9:45 P.M.

I asked the hostess where she got the ceramic horse on top of your stack of white books with bronze lettering and she just said, "Europe." As in "I'm not even going to get into what country I was in because you don't look intelligent enough to know about the names of countries in fucking Europe."

I bet it's — do you mind if I pick this up?

— biiitttch, it's from Target.

— so then I said, Do you even *have* a library card because if you did you could donate all those goddamn design books and you could go look at them however you want — I mean whenever you want — and then you would be also doing a service to the public. And that's what I do. I'm a PUBLIC SERVANT. I fucking serve the public. Serve it up. Whatever they want, if they want to repair their car or get into Wicca or sing "Love Me Tender" to me over the phone, I'm like: go 'head.

And you can sit there with your pink books and your European Target horses but what I want to know is . . . hold on . . . what I want to know is. Do you ever get touched? You need to get read! Not dusted! And also, let's just be honest here. Where are you hiding your, like, your Janet Evanovich paperbacks? Huh? Where's Stephanie Plum? Where's your self-help books that your owner bought on Amazon when she was upset and a little stoned and felt like her neighbor's Wi-Fi being temporarily unlocked was a sign that I had to get my life together. I mean *she.*

Oh, I'm so sorry, I guess I'm talking out loud. Excuuuuuse me for talking to a book-shelf.

I'm so glad I didn't get invited to this party.

The Morning After
Dear Fancy Bookshelf,

I apologize for getting angry and re-arranging your books into nonuniform colors and also for putting your ceramic animal decor into compromising positions on top of *Animal Farm.* I was asked to leave after I got caught looking under the host-ess's bed for the "real" books. My friend had to take me home and leave the fella she was macking on. Turns out I really Mr. Darcy'd it.

No Hard Feelings?
Annie

ATHLETICS--Jackson, Bob
--Motor Sports--Safety
--Spiffy

Dear *Street Biking: How to Ride to Save Your Hide,*

Listen, don't tell anyone, but you can stay around here if you want to. I wouldn't mind. I know you're older and a little scruffier than the other books. Maybe you don't use the fanciest language. Maybe you use the words "neat" and "spiffy" abundantly and nonironically and you caption your pictures with sentences like "It's a special feeling when man and machine are *both* enjoying it." That, coupled with the fact that you're a book about motorcycle safety that hasn't been updated in thirty-five years may turn some girls off. But I'm not some girls. I like a bit of danger.

It will be our little secret. Then someday I might come by and check out your chapter about sidecars, let you show me around the streets of Tucson, Arizona, circa 1980. That'd be "neat."

Readin' 'N' Ridin',
Annie

JUVENILE FICTION--Lobel, Arnold
--Spirit Animals
--Self-Help

Dear *Frog and Toad Storybook Treasury,*

I've been with a lot of books and had other characters who've imprinted themselves on my heart. But your main character Toad is my spirit animal.

For starters, we are both pear-shaped. But it's so much more than body types. Toad's very character aligns so closely with my own that I find myself going to some of your stories for advice. What would Toad do, I ask myself, when faced with a challenge. Would he go home and sleep or make some food? Either way, it's excellent counsel.

Here are a few ways your character really spoke to me:

"*Blah,* said a voice from inside the house."

"*After I put on my bathing suit, you must not look at me until I get into the water.*"

"Toad tripped over a rock. He bumped into a tree. He fell in a hole."

"*Frog,* said Toad, *let us eat one very last cookie, and then we will stop.* Frog and Toad ate one very last cookie. *We must stop eating!* cried Toad as he ate another."

"Toad ran home. He made sandwiches."

"*I am worried,* said Toad."

"*I will do it tomorrow,* said Toad."

"Then Toad fell asleep."

It's almost like you're writing ABOUT me. I know you're usually into younger readers, but I really think that you and I are soul mates. You bring such joy into my life. Something about you makes me want to ignore all of my adult responsibilities and sit around having tea and cake, waiting for the mail to come, which, every once in a

while, I can do. I want to live in a tiny forest house and take walks with a best friend
who is always kind to me, even though he is
obviously way smarter and less erratic than
I am. And really, minus the forest, I do. For
all his worrying, Toad is living the dream.
And actually, when I think about it, so am
I.

Dammit, Swami *Frog and Toad,* you did it
again! F the Self-Help section. You're where
it's at! How have you never been on Oprah's
book list? Do you see how we're on the
same wavelength? It's undeniable. We're
made for each other.

Lots of Love,
Annie

BIOGRAPHY AND HISTORY--NORTH AMERICA--U.S.--ROOSEVELT ADMINISTRATION--EISENHOWER TO CLINTON--JACK KENNEDY
--Whew

Dear *Killing Kennedy* and *Killing Lincoln* and just all of the *Killing*s by "Bill O'Reilly,"

Don't get your panties in a twist — we aren't getting rid of all of you — just some of your duplicate copies. We'll never discard you. You guys and the Mitch Albom books have dominated the nonfiction bestseller lists for years and show no signs of slowing down. Sometimes, I fantasize about putting you and Mitch together, throwing up a "Things Aren't What They Used to Be" sign, and being done with it. With an end-cap display of authors under sixty called "Kids These Days: The Worst."

I'm getting off topic. This note is just to

say you're strong-arming the biography section and taking up a lot of space, so we have to remove a few of you. The Kennedy section, especially, resents you a bit. All those book-spine pictures of Kennedys frolicking on the lawn and/or looking pensively over their shoulders seem to be willfully ignoring you. There is a copy of *Letters to Jackie* directly above your place on the shelf so that Jackie O's picture is — there's no other way to say this — looking down at you.

You're upsetting the patrons as well, or rather, I am upsetting them by not putting all of the *Killing*s next to each other. But I must stand firm. The biography section is organized in alphabetical order by the *subject* of the book, not the author. Yes, even if the author's name appears in a larger font than said biography subject. If we did it for you, we'd have to do it for everyone and then there would just be panic. Where would the Suzanne Somers books go? Health and Beauty? Memoir? Poetry*? Isaac Asimov has written a book for EVERY SECTION OF THE DEWEY SYSTEM, *Killing*. We can't open ourselves up to that kind of chaos in the stacks. Next thing you

* Suzanne Somers wrote a book of poetry in 1973 called (dramatic pause) *Touch Me*.

know, we're alphabetizing the entire collection by title, which is a proposition we receive in the suggestion box at least once a year.

No. Better that you remain next to *Johnny We Hardly Knew Ye* and just stop riling people up. Don't worry about your extra copies. They will get snapped up at the book sale. I do wish the best for you, unless the best is more famous political figures dying.

Peace,
Annie

FICTION--Mosley, Walter
--Detectives, L.A.
--Blue Dress, Devil in A

Dear Easy Rawlins Mystery Series by Walter Mosley,

This is all moving so fast. It isn't like me. I don't even like mystery series. Somehow, though, I just can't get enough of you.

God, I love wandering down dicey alleys with you, roughing people up if they need it, experiencing two decades of Los Angeles, the '40s through the '60s in all their jazzy, turned riotous, turned flower-child grandeur. Cruising down the strip when I should be in bed with nothing but my favorite PI, Easy, and the desert air on my face. But seriously, I should be in bed. I have to be up by six tomorrow.

My infatuation with you, paired with my fanatic *Moonlighting* viewing (more private eye-ing, this time shitty ones in '80s L.A., with young hunk Bruce Willis, so I'm getting the whole range of the genre), is seeping into my real life. The one where I'm a Midwestern mom burning oatmeal and forgetting to pay bills on time because I stayed up all night reading (and watching) crime noirs.

What's worse is that I've begun thinking of myself as one of your characters. I've started wearing dark sunglasses and popping my collar when I'm in public, looking furtively over my shoulder every few minutes, which makes the other playgroup moms nervous. I can tell. They've stopped offering me their extra Luna bars.

When my oldest sister called to ask if I was coming to Thanksgiving, I sniffed at her and said, "Who wants to know?" And the other day I tucked a dollar bill in my coworker's shirt pocket while telling him I'd sure be interested to know if there were any hammie sammies left after Gene's retirement party. You make sneaking and subtle bribes look easy for Easy, but I make it look . . . unstable.

If you could mellow out on the thrills, the dark dialogue, the fascinating historical details, I could handle it. But it's too much. As much as I'd like to, I don't have time to lie back with a stiff drink and fly through all fourteen of you at once. I've got a toddler to wake up with and laundry up the ass, and then also I have to figure out on what board to hide all my Bruce Willis pictures on Pinterest so no one discovers my disconcertingly abiding love for him. It's my own hard luck life, but somebody's got to live it.

Until I catch up on my breath, my bills, my sleep, I've got to leave you behind and go underground. I'll be your latest missing persons case.

Signed,

(See how I left it blank there, like I'm already gone? I've picked up a few pointers, eh, ehhhhhhh?)

AMUSEMENT AND RECREATION--
Cosgrove, Frances
--The Stage
--Thespians, Am I Right?

Dear *Scenes for Student Actors,*

The scene is a public library. ANNIE sits next to a toppling pile of books. She is a charming, attractive, and well-poised librarian. She picks up the first book in a series, SCENES FOR STUDENT ACTORS, a voluble set of volumes, to say the least.

ANNIE: Can we talk?

SFSA (*criminal with Italian accent*): ". . . Just let's talk, what do you say? I haven't talked to anybody so long I'm full of gab."

ANNIE: Okay. You see — you've been here — quite some time NOW —

SFSA (*bending over and smelling the*

75

ham): "Pshaw, Deenie, there's no harm in that is there?"

ANNIE: It's Annie, actually. We think you're great. But you haven't been checked out in ten years. I know the Theater section moves a little slower in general —

SFSA: "You see, it's not an easy job selling to people who want to get out of the shop without buying anything."

ANNIE: Well, that's kind of the deal here. Everything's free. People can take whatever they want home with them. But they don't seem to want to take you. (*pregnant pause*) Ever.

SFSA (*with sudden passion*): "[. . .] No. It ain't that! It's sump'n else. I don't understand it. I'm afeard. I'm too young."

ANNIE: You are eighty years old.

SFSA: " . . . she wasn't no kid, like the DA said she was. She was over sixteen. You seen her pictures."

ANNIE: Yeah. Anyway, so since you're old, we need to get some new stuff for the student actors to read. So —

SFSA: "You daren't look me in the eyes [. . .] D'you think I don't know why? You — a gentleman! Insolence, ignorance, and dirt! Your sport the cockpit and bearpit, gambling, and obscenity, making a beast of yourself with drink and debauching."

ANNIE (*incensed*): Hey! There is NO need to bring up my drinking. I'm just trying to —

SFSA: "My moral art avoids your tidy little mind altogether until it has fed your blood."

ANNIE: Jesus, you're taking this really personally.

SFSA (*bitingly*): "That's just it. Here I am the one person who could save this House and you bind my hands. You couldn't condemn a man to such a purgatory, Morell."

ANNIE: My name is ANNIE. And if you would just listen to what I'm trying to tell you —

SFSA: "This pain! . . . this pain! . . . All the words in the world cannot appease this pain! (*louder*) This pain! (*crying out*) This Pain! . . . There are flames, there is a fire within me! . . . Where then . . . where is the means to put it out?"

ANNIE: I hate weeding the thespians.

SFSA: "Say it ain't so, dearie, say it ain't so. (*is now vigorously powdering her ample bosom again*)"

ANNIE: It's not so! Just be quiet for a minute and let me —

SFSA: "You've done a terrible thing: you opened my eyes and heart — and you never touched me. It hurt — every bit of it hurt — how could it not hurt, it was so beautiful!"

ANNIE: All right, you made me do this: SHHHHHH! We're not getting rid of you! We just bought some updated plays and we have to shift you to make room. That's what I've been trying to say.

SFSA: "That's the nicest thing you ever said to me, Mary."

ANNIE: My. Name. Is. Annie.

SFSA: "I suppose you feel kind of lost . . . You mustn't feel lost, though. I mean, that will wear off. Life isn't going to be empty from now on. It's going to be fuller than ever! And richer! For both of us, Sam! Think!"

ANNIE (*fatigued*): You know what? Stay here. I don't care. I just want to get the hell out of the 792s and move on with my life.

SFSA: "Still I don't know. Sometimes, I think maybe it's just the way I play the saxophone. Tell me, it isn't that, Bill. Tell me, I've got to know. Tell me, it

78

isn't just the way I play the saxophone?"

ANNIE: That makes absolutely no sense. (*Exits.*)

SFSA: "Hello? Hello? (*sniffling* . . .) She — she never considers me. It never occurred to her to kiss me goodbye."

HOME ECONOMICS--Jackson, Carole
--Clothing, Fashion, and Relationships
--Colors!

Dear *Color Me Beautiful,*

Maybe there are other people who've forgotten about the most important revolution of the '80s, but I remember. The Color Revolution. As in, *doing* your colors. As in, "Banish anything navy in this house immediately! We're dyeing everything that ends up next to my face deep rose! I've just discovered — I'm a Summer!"

Even my mother can't remember and she's the one who bought you. (She's a Spring. PS: but you've got her pencil markings so you already knew that.) But *I'll* never forget. My beauty ideals are rooted in between your pages.

My mom and sisters didn't keep magazines around the house when I was little. We all wore hand-me-downs from the ladies in my dad's office, who knew he had four daughters and piled the back of our station wagon with garbage bags full of pantsuits and "silk-like" blouses. As a young child, when I wasn't wearing a forty-year-old woman's suit jacket and the Save the Rainforest culottes my mom made me (special occasion), I could usually be found in an oversize *Boblo Island* T-shirt with no pants (casual). What I'm trying to say is that I didn't have a ton of style inspiration at my fingertips. I coveted the hair accessories on *Kids Incorporated,* but we only got the Disney Channel on free weekends. It was hard to keep up with trends.

But there was you. With your color photos and charts, quizzes and makeup tips. I sat for hours with you determining my Color History. It actually doesn't take that long to evaluate your history when you're seven, but I was still sounding out words back then. I came to a sure and enduring conclusion: I'm a Winter. Just like Sally Field. How glorious.

We've grown older together. And I can tell you this. In the twenty-odd years since I've

81

read you, I have NEVER worn tan or terra-cotta red. That's loyalty. Could your makeup suggestions use an update? Sure. There's not a lot of smoked-teal eye shadow happening today. But your revolution remains relevant. I've noticed that J-Lo only wears, like, four different colors — she's an Autumn. She may not say it, but you know she's got a copy of you somewhere in a vault with her vials of goat-placenta potions and diamond cream.

So age be damned! You're my beauty mentor and I'll never give you up. It's in my will that the mortician has to consult you for my blush shade (magenta) when I kick the bucket. It probably goes without saying, you've colored *me* beautiful.

Stay Golden (unless you're supposed to
wear silver tones),
Annie

JUVENILE FICTION--Langton, Jane

--Books, Magic Inside Of

Dear *The Fledgling,*

You did a good thing for me. You did several good things way back when I first read you. But it wasn't until later on that you did the Real Good Thing.

As a kid, you reflected my love of nature with your spindly, stoic young heroine, Georgie, who sneaked away to her house of bushes, hidden from the world. She made dinners of grass salad and stone pudding and poured imaginary tea into acorn tea-cups. You held wonder in you as you told of Georgie learning how to fly with the help of a magical Goose Prince. You introduced me to Henry Thoreau through your story, which is set in the famous philosopher's hometown of Concord, in a house that invites students to learn transcendental knowledge. The interest I had in Thoreau's ideas began to bloom with you, and grew into deep admiration, until it rooted itself in my heart with such a strong hold that I named my son Walden, after Thoreau's masterwork.

That's when you gave me the Real Good Thing. I returned home, after a week at the hospital with a tiny child, and nothing prepared for him but a stellar bookshelf of my own favorite reading. My broken, leaking body didn't feel like my own anymore. My mind didn't feel familiar to me. I cried and I cried, and everyone told me not to worry so much about the baby. And I was ashamed to tell them that I was crying with worry for myself. I was at once more alone than I'd ever felt, and never going to be alone again.

You might not have known it, but among the marvels inside of you — the flying children, whimsical transcendental scholars, and talking geese — you gave me a magical portal that led me back to myself. One early morning, as I held my sleeping son, too nervous to stand and place him back in his crib lest I wake him, I grabbed the closest book within reach and was again pulled into your world. And for the first time since coming home, I felt not like a sleepless frantic stranger, but like *me,* the same me who sat on my back porch steps with my feet in the long grass, reading you for the first time. I remembered how I'd giggled at old Miss Prawn's plastic flowers, and the formation of the Georgie Protection Society. I'd swelled with hope each time Georgie leaped into the air. These memories helped me untangle from my anxious mind. You comforted me in a way that no one else's words could have managed, reminding me of my own natural soul. Of the person I am when I don't have to be anything else.

When people say books are full of wonder, we don't take it seriously enough. You are over thirty-five years old. You smell like old paper and smudged fingertips. You've lain dusty and untouched for decades. And you're magic. You are. You can't work wonders for everyone because, like all things with magic inside them, you have to wait for the right hands to touch you at just the right moment. But when it happens, it is with as much power as any ruby slippers or prince's kiss.

All through your pages, the Goose Prince speaks of a present for Georgie. In the end, the Present appears to be an ordinary rubber ball. Except that when the little girl is alone, it comes alive. The ball glows and spins in her hands, and she realizes: it's the world. Georgie knows then what the Goose Prince meant when he asked her to take care of it, and she knows that she's capable of keeping that promise.

I finished reading you and placed you back on your shelf, but your magic still glimmered around me. I looked into my arms at my sleeping Walden. I saw the girl who ran inside to look up "transcendental" in her mother's encyclopedia. I saw a boy with my eyes sinking his feet into the long grass. I saw that I was holding my world.

Thank You,
Annie

**HOME ECONOMICS--FOOD AND DRINK
--FOOD HISTORY--Williams,
Barbara**
--Popcorn
--More Popcorn
--Even More Popcorn

Dear *Cornzapoppin'*,

Hey Hay Corn-ay. You're a great little book. You've got a snazzy title. You're dedicated to someone's nonie, which is sweet. It's obvious that you love popcorn and that you've spent a lot — A LOT — of time coming up with popcorn recipes for every holiday. Even Flag Day! You are delightful and you're going to make a swell book — for someone else. At the used book sale.

I love popcorn. I guess I just don't lahooooove popcorn. But it's sort of all you ever want to talk about. Popcorn fondue, popcorn castles, popcorn lollipops and snowmen, popcorn Easter baskets filled with popcorn Easter eggs. On the Fourth of July, you suggested shooting popcorn out of cannons. When I mentioned my father was watching his weight, you recommended your "Beefy Popcorn Alternate for Dieting Dads." I don't know what kind of diet books you've been shelved next to, Corny, but most of them frown upon throwing a stick of butter and a jar of beef jerky on top of a bowl of . . . you guessed it: popcorn.

Look, you've been here since 1976. That's a long time. Not as long as popcorn, which — you informed me — may have been at the first Thanksgiving dinner. But long enough that I'm confident everyone in town who's shown interest in making popcorn cornucopias has had their shot.

It might be nice going somewhere different. Maybe someone's nonie will pick you up at the book sale and take you home! Or maybe you'll just have some alone time on the table to think about other snacks. Peanuts, maybe. Or just regular corn, if you're taking baby steps.

All right, Corny. I "butter" be going.

Stay A-maize-ing,
Annie

Dear *The Namesake,*

This is unfair. I'm on a date night, sitting at the table next to you, several copies of you actually, each with a pair of hands politely folded on top of your cover. You're at book group. And you need help.

If someone was choking, the waitstaff would call out "Is there a doctor nearby?" No one has ever yelled "Is anyone here a librarian?" at a busy restaurant. Even though these folks are clearly struggling with their grasp of you.

Most of them liked you, that's all well and good. But no one can quite say *what* they liked about you. They say, "It was a nice story." They titter, "Sometimes, the food they talked about made me hungry!" How desperately I want to lean over and say to these well-meaning ladies, "I'm a professional. I'll take it from here." My husband, Michael, has offered me ten bucks to do just that, on the condition that I use the voice of the Swedish chef from *The Muppets.* He doesn't know my anguish.

If I were in charge, we'd start with a discussion of names, of course — their meanings, their importance, why Gogol hated his. Why did Gogol's father keep the meaning behind it a secret? I would ask the group to tell me their favorite nicknames — that would open them up a bit more. I like to go off topic at book talks to build an intimacy. Then people don't feel so uptight about their opinions.

We'd segue into talking about place. How are America and India contrasted in the story? How did each character struggle between the two cultures? How was each character's story specific to the immigrant experience and how was it universal?

What about your language, *Namesake?* The easy flow of your story? I overhear one woman complain that nothing much happened. Is that true? Does it matter? Do they prefer beautiful writing and commonplace plot over an action-packed story with a broken voice? I want to know!!! They haven't even opened you yet.

"Talk about Gogol's lovers," I whisper into my wineglass. I cough loudly as I say: "Howdidhisfamilialobligationsaffecthisrelationships?" Michael's brow indicates he has gone from amused to annoyed.

I have to remind myself, once a book is written, it's in the hands and minds and hearts of the reader. I do believe that. Whatever you did or didn't do for the people in this group is their own affair and no impeccably designed book discussion is going to eclipse a reader's original, visceral reaction. After tonight, you'll be passed on to siblings, spouses, children, and friends or go back to your home library. It's fitting that you journey from reader to reader, looking for a comfortable fit. You could be met with ambivalence or you could make someone hungry for samosas or you could redefine a life. Rename it, one might even say. See what I did there? God, I'm good at book talks.

Happy Travels,
Annie

CULTURE AND INSTITUTIONS--
SEXUALITY--Cohen, Richard
`--Some Bullshit`

Dear *Coming Out Straight: Understanding and Healing Homosexuality,*

I've been waiting a long time to say this: it's over.

No one has checked you out for five years. That means we're allowed to dump you. And I do mean dump. You're going in the recycling bin and you won't be "coming out."

Guess how I found you? We were pulling books for a display about same-sex marriage being legalized! Yay, right? Oh, sorry.

Actually, you predicted that gay activists were going to demand marriage rights. But you also said that the sins of our distant ancestors might make us homosexual, so you're kind of a wild card, eh?

I think you need some space away from . . . everyone. To figure out who you are. Maybe practice some of that therapy you're always preaching. Except the Gesture Re-Education stuff. You were joking about that part, right?

Oh.

Sorry.

(Not Really),
Annie

BOOKS--Rarities
--Obsessed, No I'm Not. Shut
Up.

Dear Book That Jeffrey Eugenides May
Have Owned and Written Personal Notes
In,

I didn't move to my favorite author's hometown *because* I was looking for you, but I
can't say that it doesn't cross my mind when
I'm at used bookshops and garage sales. Will
you be there? Will you be a tattered *The Ae-
neid* or a creased and dog-eared Philip Roth
(yah, I looked up his favorite books, okay)?
Will the copy of *Anna Karenina* that I finally
read be one that he scrawled "Jeff" on the
inside cover of? Does Jeffrey Eugenides
scrawl or does the movement of his hand
follow the lyrical flow of his mind? Does he
jot? He seems like a jotter.

Maybe you'll be something totally unexpected, like a Meg Cabot novel or *The Zone* or Russell Brand's *My BookyWook*. And I just keep hoping you're going to have a secret note inside that everyone else at the yard sale somehow missed. Something to give me insight into his process, like:

"How many suicides can I fit in a book before I have to put it in the title?"

"Why isn't anybody writing about intersex people???"

"Note to myself, Jeffrey Eugenides: that girl who keeps harassing my agent over e-mail about her thesis project sounds like an intelligent and complex English major. Maybe use her as muse for character in book about a female English major?"

That would be great.

I'll Be Thinking of You
(in a totally chill, not obsessive way),
Annie

FICTION--Tolkien, J. R. R.
--Hobbits, Yay!
--Adventure, Meh

Dear *The Hobbit,*

I don't remember how I acquired you. You've always been there on my shelf, which was magical and gave me good feelings about you. So when my nephew told me he was reading you, I thought he and I could read together and have a mini–book group about it. Bonding!

I loved your beginning. Hobbits are pretty boss. They love to eat and drink. They're quick to laugh, but also get salty when they're out of their comfort zone. They rock the bright colors. I wanted to spend some time getting to know these free-wheelin', big-footed folks.

But when the main hobbit dude, Bilbo, leaves his cozy hobbit house to go adventuring with the dwarves and the wizard, my attention began to drift. This was around chapter 2. I put you aside, not worrying about keeping pace with my nephew. I mean, I can read faster than a ten-year-old. Or so I thought. Two days later, he called and wanted to talk about Thorin, and the Battle of the Five Armies, and all that other shit that did NOT take place where the hobbits live. Then I had to cram. And I finished you. But the story didn't go where I wanted it to. Well, it did — back to Hobbiton. But then it just ended. And I was still waiting for more hobbits!

I get that a lot of people love you and you "changed fantasy forever," or whatever. But if it were up to me? Maybe have Bilbo hang out with the other hobbits for a little longer. Smoke some pipe-weed. Get into trouble. Find a lady friend. Make merriment. That's *The Hobbit* I was looking for.

You are a brave and clever and longer-than-you-seem book. We just want different things.

Off You Go,
Annie

JUVENILE FICTION--Dahl, Roald
--Kids, Precocious
--Books, World's Best

Dear *Matilda,*

I've wanted to write you for so long. Since I was just a kid — before I had the right words to tell you how much I loved your dark humor, or thank you for making a bookish girl with DIY bangs like me the hero of a story, or tell you that I still think of Hortensia every time I see someone mowing down a bag of greasy "potato crisps." I've always wanted to write to you, but I never did. The truth is, I grew up. And now that I'm finally taking the time, I'm afraid this must be a very different kind of letter.

It's time for us to part, Matilda. Someone stole a few of your pages. Or, more likely, they fell out — you are frail and your complexion has darkened with age. The part of you missing includes the introduction of Miss Trunchbull, an essential piece in your narrative. Without all of your story held together, you're not making much sense.

Maybe your pages have slipped behind a bedside table or been muddied on the floor of a school bus. You've been inside the backpacks and under the beds and in the grubby hands of over fifty "precocious" children, as Miss Honey would say. I wish I could give you to fifty more, but it seems we've run our course.

Taking you out of the library is like s'm-lesses. Do you know what s'm-lesses are? It's what my mom called it when we couldn't afford to put the chocolate on s'mores. That's right. Before I met you, I was just some poor kid eating generic graham crackers and marshmallows like that's even a real snack, and also being forbidden to watch *Pee-Wee's Playhouse* because Pee-Wee got caught at a nudie theater. My folks were at their Wormwood-iest.

And then I found you. My first big chapter book. I saw myself in you. I wasn't spunky like Pippi Longstocking or mischievous like Ramona Quimby. I was shy and shabby, with my head in the clouds and in the books. A Matilda. You made me feel like anything was possible. When I was with you, I felt like I could move things with my mind! Yes, I strayed to other books and maybe left you lonely. Spontaneous thrills with *The Boxcar Children.* Talking dirty with Judy Blume. But I always loved you, *Matilda.* I never wanted it to be this way.

There will be other copies of you here, of course. But none with the original cover like you. The others have brighter colors and Matilda is standing on top of the books or waving her arms, demanding attention. Not sitting quietly, thoughtfully scratching her chin and contemplating how to punk her parents.

For me, it won't be the same if your cover isn't a little bit raggedy, your pages crumpled, an illustration of Mr. Wormwood's checkered jacket outlined in wobbly pen. The library will be a little different without you. And I'm a little different because of you. And I'm grateful for it.

Sincerely,
Annie

FICTION--Steinbeck, John

Dear *Cannery Row,*

What are you doing here? The back of a
toilet is no place for you, an American classic, the respected midcareer work of a Nobel
Prize winner. Not even a cruddy wire shelf
next to the toilet — the sweaty top of the
tank itself, strewn with stray hairs and the
crusty warts of what one can only hope is
dried hair gel.

You're not alone, but you're not in suit-
able company. If you were in a tidy arrangement of other titles all referring to "the can"
in some way, that would be clever. But
instead you're topping a sad lumpy heap of
reading material, reflecting the literary
tastes of all the roommates who live here: a
wrinkled *Musician's Friend* magazine, a

bloated (from water damage, be nice) copy of the newest Jonathan Franzen, a guide to home brewing beer, and an immaculate *Money Book for the Young, Fabulous and Broke* by Suze Orman, with the inscription, "She knows her shit! Love, Mom, Christmas 2005." Is it good to be on top, *Cannery*, even with this bunch? I guess the less fluid your poor worn paper cover is soaking up, the better.

I get why you're here. You're a quick read, you're about the everyman, you revel in uncomplicated pleasures, one of which, okay, is shutting the door and reading as much as you can of a good book before your legs go numb.

Some of your main characters live in a vacant fish-meal warehouse, uncannily like this particular bathroom. The word "stink" is in the second line of the book as well, so I understand you're no stranger to your surroundings. But, please believe me, even though one of your major themes is to respect and find wisdom in the castoffs of society, to not place yourself above the petty thieves and the prostitutes: YOU ARE BETTER THAN THIS BATHROOM.

I know these guys. I love these guys. But this place is floppier than the Palace Flophouse. Your quote "no money, and no ambitions beyond food, drink, and contentment" couldn't fit a group of people better if you'd *Quantum Leap*ed to this very time and location from 1945 and eavesdropped on a conversation about the flavor nuances of the Taco Bell breakfast menu. The party banter may lean philosophical, but the probability of leaving with pink eye is TANGIBLE. For your safety, we've got to get you out of here.

Though you're weathered, you've still got that quiet lopsided charm about you. I even thought about smelling you. Your pages look just the right shade of yellow, your text just dark and smoodgie enough to give me a deepnostril thrill ride. The only better smell than a creaky paperback for a book sniffer like me is a real inky graphic novel. I mean, I'm not *going* to smell you. But I am going to get you out of here before your owners inevitably forget to buy toilet paper and things get desperate.

Run away with me. We live in modern times. They can look at their phones when they're taking a dump like everybody else. I've got baby wipes in my purse. Maybe I'll come back for the Franzen, but I gotta go now. My legs are going numb.

Smell Ya Later,
Annie

ECONOMICS--Herera, Sue
--Finance and Investing
--Boooooorrrrring

Dear *Women of the Street,*

To be honest I thought you were going to be about hookers. I have an old book about band groupies of the sixties. It's appropriately titled *Groupies.* And it's got interviews with all these women and men with fun names like Silky and Sunshine and Pogo. They ball their way through the British Invasion and have great hair. But they also speak frankly about how lonely the life is. You really get to know their crazy asses. I don't want to compare you to other books, but that's what I thought you were going to be like.

Sigh.

You're about finance. You interview

women named Muriel and Linda and Bernadette. No one has good hair, even though the illustrations are charcoal drawings. We found a business card advertising a weird and expensive-sounding plastic surgery in you when we pulled you off the shelf. And it's been in there since the '90s.

Your book jacket said you were provocative. I thought you would be out on the Street all the time. But you've been checked out twice. Like twelve years ago. A lot has happened out there while you've been in here on the Shelf. For starters, your profiles of Goldman Sachs employees need tweaking.

All right, I'm being snarky now and that's not fair. It's not that I *wanted* you to be about hookers. Finance can be a fascinating topic (Probably. I know all sorts of people come in looking for the book by that *Free Money* guy with the question mark jacket), and I'm sure these women worked really hard. So why do you have to refer to the longevity of their careers as the amount of time they've "been on the Street"?

Is it a Wall Street thing? Or did you maaaaybe sort of want to make people think about prostitutes? World's toughest business . . . world's oldest profession. Is it just me? Maybe it's just me. You're just not who

I thought you were.

   See You at the Book Sale,

     Annie

FICTION--Maguire, Gregory
--Witch, Wicked
--Bitch, Still A

Dear *Wicked: The Life and Times of the Wicked Witch of the West,*

You were a sure thing, with your gilded pages and your fancy map on the inside cover. I can't resist a book with a map in it. The combination of my love for the film *The Wizard of Oz* and the children's book *The True Story of the Three Little Pigs* by Mr. A. Wolf (and Jon Scieszka) made me feel as if we couldn't go wrong. I felt a special, singular connection to you, but I also felt camaraderie with all the people who were simultaneously realizing that they *had* to know why the Wicked Witch of the West turned out so rotten and green. It seemed like a good idea at the time.

So I bought you. And I slugged away at your four-hundred-and-then-some pages for months. I put my all into our relationship. I wanted to love you. I wanted to gush with my sister about you. I wanted to curl up on the couch on Thanksgiving and watch *The Wizard of Oz* from a different emotional vista. I wanted to see Elphaba the witch as more than just a bitter crone with a shoe fetish and really terrifying monkey friend-servants. I don't know if you can understand the intense glee and expectation that make up the book lover's baggage when approaching a tale like yours, a book that should have been there all along, and now suddenly was in my hands.

But after copious, co-pi-ous explanations of Ozian religious and political regimes and animal rights and looooots of misplaced resentment, I still think the western witch is a spiteful stone-cold bitch. And I get that the slippers were, like, a representation of her father's love for her sister and not for her. But, come on, give it up, Elphaba. You were born green, big deal. Your sister had NO ARMS! She coped. Until that house fell on her.

We should have parted ways "In the Vinkus," but I had a weird thing about staying loyal to you until the end. And like all relationships that last longer than they should, I let my anger boil over. I still couldn't connect with your main character, and, on top of that, I was pissed about the other books I could have been reading. I had the urge to douse you with water and see if you evaporated before me like the old witch. So this is it. I'm putting you in a Little Free Library.

You're Dead to Me,
Annie

Dear Nikki Giovanni's *Love Poems,*

Oh geez. How can I describe love better than you? Can't be done. So I have to try and stutter your own words back to you, like a . . . I'm no good at this . . . like a girl with a crush on her first-period Spanish teacher? Like an old car engine in the winter? Baach, I'll just leave it up to you.

When I pick you up, I feel "glad as mortar / on a brick that knows / another brick is coming" because I know how much satisfaction I'm about to get. I'm about to get warm and smiley all over. You are everything love is. You're sexy and playful and dizzy and also powerful and vibrant. And sometimes hurt and wanting more love.

I like to have you around in the kitchen so

I can read a poem while the water boils and another while the butter melts, and so on. It's a reminder to read slow and savor you, and the smells of the cooking make me more aware of my senses. "I Wrote a Good Omelet" is best read while frying eggs with no pants and bedhead. I know you know what I'm saying.

And I like to have you around on the coffee table so I can show you to my friends and we can talk about "And Yeah . . . This Is a Love Poem" or declare "Love Is" to one another.

But other times I'd just like to be alone with you so we can get into "When Gamble and Huff Ruled" and "Seduction" with no distractions. You're a skinny-looking book, but you're so big on feeling. Days after I've held you, I'm still thinking of you. Also you mention "Slow Hand" by the Pointer Sisters.

So,
you are
it
for me.

(That's a poem for you.)
I really like having you around. I'm so glad

you exist.

LOVE,
Annie

FICTION--McCarthy, Cormac
--Shush Now

Dear *Blood Meridian,*

You had one job. Keep him quiet. All your bloody bravado, and you couldn't even manage to silence one husband for one goddamn night while I finished my book.

I had the idea last week when I wanted to read and Michael wanted to watch *Breaking Bad.* I thought, "Hey, this guy seems to be entertained by death spirals of unrelenting hopelessness. Maybe if I found a book that matched those sensibilities, he could enjoy that same broken-human trauma noiselessly, and I could read my twee novel in peace."

Enter you.

It was a brilliant plan. Ah, but it worked too well. Have I read you before? No, sir.

But I know all about your southwestern militias and your dead-baby trees and your Apache scalps. I know it all, compadre. I know It All. Because my husband won't stop talking about you.

Here I am, trying to read my quirky Miriam Toews novel. It's set in the smallest town in Canada. One of the characters is named Summer Feelin'. I have a perpetual "oh goody" smile on my face while I'm holding it. Then every five minutes:

Michael: "Shoosh. Men can be brutal."

Me: "Mm."

Michael, three minutes later: "This writer really tells shit how it happened, man. He's like a war correspondent for a war that happened a hundred years before he was born."

Me: (aggravated sigh).

Michael, five minutes later: "The judge dude is the fucking devil, man. Let me read you one more part."

Listen, I'm glad he likes you. I'm less than charmed with the philosophical theory you're advancing that "humans are still awful and selfish; we've just come up with ways to fuck each other up without scalping." But I was willing to deal with even this, if it allowed me to find out whether small-town Mayor Funk gets his visit from the Canadian prime minister after all. At the conclusion of MY book.

Now I'm confusing plots. *Where was I? Oh right, It's the Canada Day festival and the Mexican militia shows up and takes the Kid and the prime minister as slaves. No . . . Summer Feelin' is trapped in an outhouse with the judge devil. Wait.*

In order to avoid my becoming violent with my husband, thus proving his burgeoning theory, could you tone down the mind-blowing prose and the blind carnage? I've got like twenty pages left.

Take It Easy, Will Ya?
Annie

Dear *Yertle the Turtle and Other Stories,*

I think we're going to be friends. Not you and me — we're already tight. I love your wise tortoise rhymes. I mean that I think I'm going to be friends with your owner. She's a new mom I met at the park and I'm here on my first playdate. Seeing you sticking out of her daughter's diaper bag, your tale of ne'er-do-well Yertle and the toppling throne of turtles defeated by one burping activist tortoise called Mack — that says to me: this lady and her young child are probably not dicks.

You couldn't have shown yourself at a better time. If I don't find someone to buddy up with at the park, I'll have to stay in my "assigned seat" on the outcast bench, with Bluetooth dad and the grandma with the

MARGARITAVILLE T-shirt and the smoker's cough. In the caste system of the Tot Lot, we're down in the mud like your friend Mack there. We are without the kingdom comforts of jogging strollers, muffins, and discount codes for essential oils.

Those other moms, they are the rulers of all that they see. And what they see are other people's weaker-than-thou kids taking too long at the slide and going barefoot in the community sandbox like goddamn heathens. It's no wonder their own offspring feel the need to push the underlings off the top of the climby thing to get by. What were they supposed to do? Wait in line like a commoner child?

Since your release in 1950, you may have assumed fascism was dead, but you need only look around this lot of tyrant tots and their proud parents to see that not everyone absorbed your line about all creatures being free. While the parents at the top of the proverbial turtle heap discuss mind-body connections and preschool plans on the shady benches, I'm down the throne on the broiling-hot-covered-in-bird-shit bench, wiping sand out of my crying child's eyes with the bottom of my T-shirt. The mother of the kid who threw the abrasive debris into my son's face shrugs to me from afar and calls out, "She loves to throw sand!"

Fuckin' Yertles.

That's why I'm so happy to see you. If this new mom friend believes in you the way I do, if we can somehow manage to disconnect that dude's Bluetooth and wake Margaritaville grandma up, I think together we could start a revolution, not unlike the one in the pond of Sala-ma-sond. If all of the "little" people get "a little bit mad," we can defeat the dictatorship going on over by the teeter-totter, the judgment of our snacks, and the unabashed direct sales pitches. I say we rise up and shake this throne!

Who's with Me?

Annie

FICTION--Bradbury, Ray
--Science Fiction
--. . . Or Is It?

Dear *Fahrenheit 451,*

Don't ever change. And stay here with us, always. You were created in a library, and I'm comforted by the fact that you'll remain on library shelves around the world. If we ever get to a point when you're not included in the core of a book collection, we're all fucked. Like "Our civilization is flinging itself to pieces. Stand back from the centrifuge" type fucked. Some days the world feels closer to that point than I'm comfortable with.

Be glad you have a voice but no eyes. Since 1953, the talking walls are bigger and louder than ever. The modern-day "firefighters" are armed not with kerosene but snarky Internet memes, reality TV, and the ability to simultaneously see more and less of the world around them. I shouldn't even tell you, but there are people who don't believe libraries are necessary anymore. A bunch of Captain Beattys. It's frightening.

Oh I know what you'll say: "We need to be really bothered once in a while. How long is it since you were really bothered? About something important, about something real?" You're right. That old noodge Ignorance plodding too close behind with its hot dumb breath on your neck can help up your game.

The library is a good gig to have — convincing people that I want to help them with literally WHATEVER they want to learn about. Helping create more "examiners, critics, knowers." But it's harder than it might seem to get people to listen.

Maybe it would be easier if we were allowed to yell? The public librarian has been typecast. We're supposed to whisper and shush, demand silence, when in reality we work our asses off trying to help people speak up. Maybe it seems safer for us to whisper. Because maybe if we could shout, it would shake the walls down.

For now I'll take it step by step: test-prep book by car repair manual by "Can you help me apply for this job?" One phone number for a shelter, one kid psyched to be handed the next Percy Jackson book, one woman with no one to lean on who needs a divorce guide, a new e-mail account, a *Small Business for Dummies,* and an encouraging smile. One book, one patron, one desk shift at a time.

I'll be quiet about it, like you, just sitting on the shelf, waiting for someone to come to you with a question she or he doesn't quite know how to ask yet. "In the silence, our stage whisper might carry," right? But don't ever go. We're in this business together.

It Was a Pleasure to *Learn,*
Annie

FICTION--García Márquez, Gabriel

--Love 'Em and Lepers

Dear *Of Love and Other Demons,*

I did something sort of rash the other day, and now I see you're on the reshelving cart, so soon after you were checked out. I just want to tell you, you didn't do anything wrong; it's all my fault.

It's my job to match books to readers. You go out often enough. You certainly have your own adoring crowd among your author's completionists, and there was really no reason for me to send you out on this particular assignment. But I did it anyway, and I'm sorry. I hope once you see the situation, you'll not take things personally — book-a-sally — booknally? Whatever. I'll start from the beginning.

A lady came in whom I know to be a fan

of paperback historical romances. And she was going on about the last book she'd read, *Captive Innocence* by Fern Michaels, which is about a Puritan New England beauty who travels to Brazil and falls in love with her neighbor who is also her worst enemy. Then this reader, she asked me if I knew of any other historical romances set in South America. And this little voice in my head was like, "Do it."

Did I know that a devoted Fern Michaels fan was perhaps not the best audience for a book about how a seventeenth-century tween who is bitten by a rabid dog and the priest who visits her in her jail cell/torture chamber fall in love? Yes, I confess I did. Could I have reasonably assumed that your first few pages, which mention crypts, slobbering beasts, tortured slaves, and disfigured dead bodies washing up on shorelines, might dissuade this particular reader from reading 120-odd pages *further* to the part where the twelve-year-old, now believed to be possessed, and the priest finally kiss after a scene where he lovingly kills her lice? Guilty as charged.

Except a little voice was goading me: *She said she wanted a historical romance set in Latin America, didn't she? Just don't bring up the lepers. It'll be fiiiiiinnnnnne.*

I guess I really wanted to "wow" my reader. If she was going to stick her toes in the Latino fiction waters, why not shove her into the deep end? A more practical librarian might have introduced her to Laura Esquivel's *Like Water for Chocolate,* then the beguiling work of Isabel Allende. Step by step. Maybe it was the sugar high from the Costco cake in the break room, but I wanted to take her all the way.

She returned you after two days. That doesn't mean anything — you're a short book. But now she won't make eye contact with me.

I don't want you to take the blame for this. You are obviously a beautifully written book and a stirring tale about romance among filth and isolation. You're passionate and funny and tragic, and your hero, the priest, crawls through a sewer to see his beloved after being condemned to a lifetime of serving lepers. That's love, man.

It just wasn't your time with that reader. There will be others, of course. But for now, it's back to the shelf for you.

Adiós, Mi Amor,
Annie

```
MEDICINE--HEALTH AND HYGIENE--
HYGIENE OF OFFSPRING--
Schwartz, Bob
--God, Swear To
--Those Are the Real Subject
Headings I Found This Book
Under
```

Dear *The One-Hour Orgasm*,

We pulled you from the shelf for review because you were on our list of books that haven't been checked out in a long time. But we're *actually* getting rid of you because your pages suggest that patrons bring you along with two large mirrors and a tub of Vaseline into a space "warm enough for you to be nude for about one hour." That's just not the kind of environment we like to encourage library readers to venture into with material that other people will subsequently touch.

I don't know who was working here in 1992 that looked in *Publishers Weekly* magazine and thought, "Oh cool, a book for people to hold in one hand while they squeeze their lubed-up balls with the other. We'll take it!" Library budgets were bigger in those days. Perhaps they could afford to replace books that were basically teabagged.

It could be that your author had a table at the ALA conference and it was post–cocktail hour and someone was coming off of a lively Collection Development seminar, so what the hell? Maybe they saw the promise to change relationships "beyond your wildest dreams" on the back cover or the unattributed blurb quotes that call you a masterpiece. Whoever bought you, either they aren't working here anymore or they're staying mum, because no one has stepped forward to claim you.

We're just not a good fit. You lead with big words like "tumescence" (spoiler alert: it's when women get mean before their periods, but actually they're just in heat); but the word you end up using most is "crotch." As in "the crotch area will be the most sensitive area of your body." Then, the first step you advise in learning how to pleasure yourself is to tidy up your house, which is where I would have checked out if you hadn't already lost me with the Vaseline business.

So it's not that you're bad, it's just that your style is a little too "interactive" for us. Not to harp on this, but sometimes books get checked out, like, eighty times before they are replaced or discarded. That's a lot of crotches.

For a single reader, though? Sure. We're not prudes over here. Working in Adult Reference, we get a good many "adult services" jokes thrown our way, and we respond to them with great poise and nary a nostril flare of disdain. This is how I know that you probably do have an audience among our patrons. And why I am certain that you must go.

All right, I'm going to go tear through our budgeted monthly supply of hand sanitizer. Try to keep your pages to yourself until I find some gloves and/or tongs.

Stay in Touch (just not touching me),
Annie

CARICATURES AND CARTOONS-- COMICS--Guisewite, Cathy --Chocolate and Feminism, Aack!

Dear *Another Saturday Night of Wild and Reckless Abandon: A Cathy Collection,*

One of my girlfriends had the nerve to roll her eyes at you as she scanned my coffee table. Oh hell no.

"Is that, like, the lady that loves chocolate?" she asked.

Well, sure, Cathy loves chocolate. She also can't control herself around donuts or guacamole. She hates how her thighs look in pastels, she lets her overbearing mother guilt her into shit, and she dates a total drip named Irving. So fucking what?

I had to give my girl a little lesson in Cathy, *Cathy.* Because when you're not talking about dating insecurities and how to eat feelings, you were one of the first to address the contradictions of the woman who's trying to "have it all." You discuss the wage gap, mansplaining, and sexual harassment. You try to explain fluid gender roles in a way Cathy's mom might understand. Yeah, Cathy has a messy room and frets over her terrible hair. She's trying to figure it out. That's what makes her so lovable. That's why she's got that heart on her sweatshirt all the time, friend. She's putting it all out there.

I'm proud to put you right by my Gloria Steinem essays and *Bad Feminist*. You may be a collection of cartoons that I occasionally cut up to mail as valentines, but you're part of the sisterhood. You're my favorite '80s woman. So, *aack* on girl.

Tiny Heart,
Annie

AUTOBIOGRAPHIES--CELEBRITY
```
--Inspiration
--Yourself, Believing In
--Juicy Bits
```

Dear Celebrity Autobiographies*,

Thank you for being so uplifting. When I read one of you guys, I really feel like I have the ability to make something of myself and maybe even snag a little fame one day, if I simply keep believing in *me.*

The typical outline for your genre is so tasty to me. Like a tiny party-snack cheese-cakey thing that you feel you can only take if you say "Oh, fine, I'll try one" beforehand, even though no one was asking you to try one. Your formula usually goes:

* Some of these are technically memoirs — don't get on my nuts about it.

- birth and home life, including a tidbit about a moment in youth that made celebrity or celebrity's relatives realize they were fated for stardom
- "those were the days" poverty while trying to break into industry
- steady course upstream in career/first marriage
- major setbacks/drugs
- redemption and humble-brag about inner strength
- recent project plug about how current work is most important thing celebrity has ever done

More or less.

I'm not saying you don't each have unique qualities! You do! But pulling one of you off the shelf and knowing you're going to hold up to that characteristic Celebrity Do Tell* formula makes the glee bubble up inside me.

Dolly's sweet nostalgia for Smoky Mountains life, a teenage Rob Lowe gallivanting with Tom Cruise and the Sheen boys, Carrie Fisher dishing the deets about *Star Wars,*

---

* I prefer "Do Tell" to "Tell All." The former sounds more like giddy friends holding hands and whispering than skeazy, invasive "Tell All." Stay off my nuts about it.

Paul Simon and electroshock therapy. You get to tell your own stories. And no matter what kind of topsy-turvy path you guys take, you always come out with unbroken faith in your self-worth and in life's ability to work out for the good.

Other reasons I love you? You have amazing subtitles. (See fun quiz on page 98!) You bring people that don't normally read into the library (here's lookin' at you, *Stori Telling* by Tori Spelling!). Your glossy picture inserts are the tops. And I enjoy scanning your acknowledgments section to see if I recognize any other celebrities by the private nicknames you give them.

Thanks for the inspiration and the juicy ex-lover stuff.

Yours Truly,
Annie

Celebrity Autobiographies: Fun with Subtitles!

Match the titles and authors to their subtitles.

Melissa Explains It All by Melissa Joan Hart

Soulacoaster by R. Kelly

The Book of Joan by Melissa Rivers

Landing on My Feet by Kerri Strug

Navel Gazing by Michael Ian Black

Ron Jeremy by Ron Jeremy

Life and Def by Russell Simmons

Bedwetter by Sarah Silverman

Tales of Mirth, Mischief, and Manipulation

The Hardest (Working) Man in Showbiz

Stories of Courage, Redemption, and Pee

Sex, Drugs, Money, and God

A Diary of Dreams

The Diary of Me

Tales from My Abnormally Normal Life

Unleash Your Inner Rock God, Win in Life and Business

The Art of Men by Kirstie Alley

Short-Term Memories of Longtime Friends

Me, Inc. by Gene Simmons

True Tales of Bodies, Mostly Mine

Make 'Em Laugh by Debbie Reynolds

I Prefer Mine Al Dente

Cash by Johnny Cash

The Autobiography

Answer Key: *Melissa Explains It All* by Melissa Joan Hart . . . *Tales from My Abnormally Normal Life; Soulacoaster* by R. Kelly . . . *The Diary* of *Me* (no explanation for the romanized "of"); *The Book of Joan* by Melissa Rivers . . . *Tales of Mirth, Mischief, and Manipulation; Landing on My Feet* by Kerri Strug . . . *A Diary of Dreams; Navel Gazing* by Michael Ian Black . . . *True Tales of Bodies, Mostly Mine; Bedwetter* by Sarah Silverman . . . *Stories of Courage, Redemption, and Pee; The Art of Men* by Kirstie Alley . . . *I Prefer Mine Al Dente; Me, Inc.* by Gene Simmons . . . *Unleash Your Inner Rock God, Win in Life and Business; Make 'Em Laugh* by Debbie Reynolds . . . *Short-Term Memories of Longtime Friends; Cash* by Johnny Cash . . . *The Autobiography* (because he's fucking Johnny Cash and he doesn't need a subtitle).

BIOGRAPHIES--CELEBRITY
--Journeys, Dark
--Nazis
--Dead Puppies
--Karate

Dear Celebrity *Biographies*,*

Thanks for keeping shit real. When I read one of you guys, I never ever *ever* want to be or know a famous person.

Your genre's tradition of going deep and getting dark is simultaneously addicting and dispiriting. Recklessly, I want to be the person that knows more interesting facts than any of my friends about the Supremes (Flo 4 Ever). But your provocative party banter and smug superiority add a leaden weight to my soul. You're like the good dip

* For a list of my All-Time Top Bios and Memoirs, see page 309.

at the party that's still too hot to eat. I don't want anyone else to get to that shit before I do, but I know that lava mystery cheese is gonna burn my tongue.

Audrey Hepburn being starved during the Occupation. Lance Armstrong shooting up his own blood in the back of a bus. Gypsy Rose Lee's mom killing puppies and throwing people out of windows. Marvin Gaye trying out for the NFL and getting weird with the hookers. You're endlessly fascinating, but you keep it so real that I sometimes can't go on. I had to take a two-year break between Peter Guralnick's Elvis biographies just to be able to read and still cope with life.

Other things I love about you: when I'm curled up on the floor after finishing you, in a weird, dark head space, I imagine the torment your authors must have gone through — spending years or even decades obsessing about the histories of their subjects but still exercising copious restraint in order to write from an unbiased viewpoint. And then I find comfort knowing that somewhere else in the world, Peter Guralnick can't sleep either because he's wondering the same thing I am: why did Elvis love karate so much?

Please don't ever come out with anything

about Mr. Rogers. It might kill me.

Warm Regards,
Annie

JUVENILE--Kirk, Ellen
--Trucks
--Trucks, Just

Dear *My Truck Book,*

Just fucking enough already. It's always the same with you. I picked you up at the used-book sale when my son was just starting to talk and I thought his fascination with dump trucks was cute. I thought this stage was going to morph into farm equipment or boats, and then hopefully something more interesting like bugs or weather or anything but more fucking trucks.

You're only six pages long, but you take a half hour to read because you point out all your parts. Your wheels and windshields and mudflaps. All that shit. And I read you like fifteen times a day. Do you know how much of my life I've spent reading you so far? I don't, because I've been too busy reading to add it all up. Also, librarians aren't that good at math.

Have you ever heard of a narrative arc? Just one time I'd like to open you and see an actual driver in one of your trucks. What's his name? What's his home life like? How high is his cholesterol? Nope. Not *My Truck Book*. It's just "this truck picks up garbage" and "this one puts out fires."

I'd cut you out of my life forever, but I know how much you mean to the kid. So here's the deal. You're going up on the shelf where he can't see you. And don't be surprised if I invite other books home with me. My son likes apples. I'm hoping that's going to be a new thing. I'm pushing for apples because they can't possibly have as many parts to identify as MORE TRUCKS.

Roger That?

Annie

SHORT STORIES AND ESSAYS--
Sedaris, David
--Holiday Merriment
--Drinking
--Dismemberment

Dear *Holidays on Ice,*

You are as snugly settled into my holiday habits as *Pfeffernüsse* cookies, Judy Garland singing "Have Yourself a Merry Little Christmas," and the traditional shot of Crown Royal taken before gift giving commences each year. In fact, you're the opening act. You on audio, terrible road conditions, and some celebratory gas station jerky unite to bring the very first shimmer of Christmas to my sanguine heart.

As one of five kids, I was awfully lonely living in Chicago and being away from my big family in Michigan as they prepared for the festivities (in the Midwest, that's mostly

141

making hot dip and uncreasing the wrapping paper you saved last year). And the five hours of slushy, icy, semi-truck-y driving that lay between me and my happy childhood home seemed like a shitty way to kick off the holiday.

I got you one year, thinking you might lift my melancholy spirits and allow me to ease my vise grip on the steering wheel while I traveled. And, old friend, you didn't disappoint.

Is it strange that a collection of essays about prostitutes and unfit Santa's helpers, stories of theater critics who skewer children's nativity plays, and families who sacrifice body parts in order to one-up the neighbors is a sacred part of my holiday season? Not if you knew my family. Which is what is so fantastic about you. You're outrageous in the most relatable way.

You are about all the things we think about (or groan about) over the Christmas season: commercialization, odd and uncomfortable traditions, and selfish people pretending to be good for the sake of the family Christmas letter to name a few. We've all been there! But you say it funnier. Your offhand delivery of shopping-mall-elf lectures that sternly remind elves to wear underwear, your casual mention of a woman who may or may not have committed murder in an attempt to frame her husband's adult Vietnamese love child. Oh, the way your character Thaddeus Bristol refers to six-year-olds as porkers and sadists! It's a good goddamn time is what it is.

But you're real sweet too. "Dinah the Christmas Whore," for example, is a beautiful tale of helping the less fortunate, and the warmth and wit in this story put a smile on my face that even the elf who wants to be a dentist in *Rudolph* can't rival.

You know what you are like? You're like rock candy. I only eat it once a year. It's got sharp edges, but it's warm and sweet if you give it time. And the powdered sugar on it makes me choke. (I guess that one doesn't fit. Maybe if the powdered sugar was, like, your irreverent humor and the choking is guffawing. There. There you have it.)

You've been with me all these years. I live closer to home now and can barely finish you before I pull in the driveway. Sometimes, I'm tempted to stay in the car with you just a little bit longer. Because you are my Christmas, *Holidays*. And you're part of the family now. Lord help us, you fit right in.

Happy Holidays,
Annie

ROMANCE--Various Authors
--Love, Smothered By

Dear Harlequin Romance Spinner Rack,

I never feel as susceptible to warts as I do when I'm weeding you guys. That's not meant as an insult, but you do get around. I mean, you're popular. Teenagers, the exceptionally old, and heavy smokers (again, mostly teens and the elderly) love you. What's not to love? You're bold. You're quick. You're to the point. No need for back-of-the-book blurbs — your titles say it all:

To Lasso a Lady
Into His Private Domain
One Night, So Pregnant!

A patron might approach me with a quandary: "I don't know what I feel like reading, but I know I like taboo affairs and Turkish men." With a glance I can zero in on *Falling for My Mediterranean Boss.* Done and done. Who's next?

You are brimming with romantic possibilities. A full rack of full racks. But that's just it. You're too full. It's claustrophobic over here. It's an orgy of *Rebel Rancher*s, *City Surgeon*s, *Billionaire Daddy*s, and *Gentle Tyrant*s.

With new titles appearing (it feels like) daily, there's not room for everyone. Plus, paperback erotica is having a moment and we need to make room for *Trailer Park Virgin,* etcetera. Now I know that's not really your scene, but if you're lucky enough to stay here, you're going to have to get along with the others. Have a safe word. That's all I'll say about that.

The rest of you will be going in a box in the breezeway with a TAKE ME HOME! sign on you. I'm not going to lie. It's no picnic in there either. You'll be rifled through and thrown into haphazard piles by folks who can remember each of the eight hundred Harlequin titles they've read but forget every Friday that we close at 5:00. EVERY Friday, dammit. Let's go, people. Get your smut and keep moving!

Listen, I know it's not ideal. But you've

had *Her Moment in the Spotlight.* I hope you can *Escape to Happiness* — somewhere else. *Love Is Fire* (I don't know what that means; it's just one of your titles), Annie

Dear *To Kill a Mockingbird,*

You're one of thousands of books I've borrowed from my big sister. You may even recognize some of your old shelf buddies here (I'm not great at the returning part). I wanted to take the time while you're visiting to say thank you.

We're all readers in my family, because my parents liked to read and they believed in education and encouraged us to always expand our knowledge. And because we were broke and didn't have cable and the library was free. And because my mother would have lost her damn mind if she hadn't come up with a way to keep all five of us quiet at the same time.

My siblings helped to mold me into a book lover in their signature ways. My sister

Renee read me bedtime stories. My sister Kristin gave me a book called *Alone with the Devil: Famous Cases of a Courtroom Psychiatrist* when I was eight years old and wanted to read "big kid" stuff, which taught me that books can have a lasting (damaging) effect on you. My brother, Adam, reveled in the art of the oral storytelling tradition, with tales about how the doctors would probably amputate my broken arm and about our fabled "other brother" Peter, who was sent away for bad behavior.

But my oldest big sister, Michelle, was the true bookworm. She had three qualities that were integral to my becoming a reader: a job at a bookstore, disposable income, and the ability to inspire fear and/or awe in me at her discretion.

She was a pop culture trendsetter in our household, which could fall largely in your favor (*Beauty and the Beast* on VHS!) or far outside it (her Michael Bolton phase). She'd come home with a new book and a direct order, "Read this." When the next *Boxcar Children* in the series was permanently marked "lost" at the school library, a fresh new copy appeared on my bed. She'd slip me five bucks at the used-book sale to fill up a paper sack of my own. John Jakes sagas for Mom; demented horror books for Kristin; a history of baseball for Dad. Even after Michelle moved out and wasn't working at the bookstore, we benefited from her insatiable reading appetite because she'd send over boxes of books she no longer had room for — all in pristine condition because she never breaks the binding or dog-ears pages.

Even now, when Michelle tells me to read a book, I do it, because you have to do what your big sister says. She was the first to tell me to read *Bridget Jones's Diary*, *Devil in the White City*, *Gone Girl*, and *The Martian*. She turned me on to John Steinbeck and Rainbow Rowell. My big sister was and remains my original librarian. Even though she could never be an actual librarian because her love for books is only matched by her disdain for the general public.

I know what you're thinking, *Mockingbird:* "What the hell does this have to do with me?" But don't you know, already? You're the one who started it all. You're the book that turned my sister into a capital-R Reader. You're the uplifting story she goes back to after she reads something shitty and has the bad book blues. Atticus is her blueprint for a good man. She quotes you when she leaves clues for hidden Easter baskets.* You're the one I have to thank for all the books I've loved since the first time she got her hands on you.

So even though you never hit me with a literary lightning bolt — the way you have for so many — you're still one of my favorite books. I don't even think I remember what you're about, to be honest. A tomboy and a guy named Boo and a court trial, right? Courage, I know you're about that. It's been so long. I read you in high school. No, I read *The Catcher in the Rye* in high school. The morning English class read you. Maybe college? OH MY GOD I'VE NEVER READ YOU.

All right, enough with the pleasantries. Let's get to reading. I've got to finish you before the next time she calls. There's Face-Time now, and she knows what I look like when I lie.

Godspeed,
Annie

* "You never really understand a person until you consider things from his point of view" means "If you sit in Dad's chair, you'll see you can see your basket behind the TV."

BOOKS--Prohibited, Lost, Imaginary
--Stoner's Delight

Dear Books I Imagine My Upstairs Neighbor Reads,

I wonder about you. It's a sort of game I play on, say, a Tuesday at 4 A.M. when I'm lying in bed listening to the last of my neighbor's party guests amble down the stairs, very likely forgetting to lock our mutual back door; when the music and what my ears perceive to be angry line dancing have ended and the microwave has stopped beeping and the phone placed on the glass-top table has been taken off the vibrate setting. Right before the last puff of dank weed smoke trails downstairs, underneath the door, into my bedroom, and the contact buzz finally allows me to relax and drift off to sleep. That's when I wonder if on that top floor, tucked into his bed with his reheated Thai food and his girlfriend with the coat Kate Hudson wore in *Almost Famous,* the dude upstairs is unwinding with a good book.

What sort of reading material might this dude have? The only other possessions I can confirm he owns are a leather jacket, a van, and an endless supply of, from the smell of my stairwell, some pretty good shit. What kind of book collection could you be? Are you the perfect stoner library?

I like to imagine a large living room coffee table, low and square, made of dark wood and covered in a tasteful, distinct selection of books, used as conversation starters among Tuesday-night rabble-rousers or the daytime visitors who double-park and leave their flashers on outside while they bound upstairs to buy pot. In my head, everyone is required to wear a forest-green smoking jacket with black-velvet lapels, except the girlfriend with the *Almost Famous* coat, who sits in the corner, sulkily reading *Super-minds: An Enquiry Into the Paranormal.* A *My So-Called Life*–era Jared Leto is there as well, tucking his hair behind his ears in a come-hither manner.

Everyone lingers around you, the book table, and you are: first, a small pile of thought-provoking reads intended to spark debates and epiphanies, like some Ta-Nehisi Coates, Susan Sontag, and *The Man Who Quit Money* by Mark Sundeen, about a guy named Suelo who gives up all his money and lives in a cave, eating watermelon from abandoned melon fields and figuring out life's mysteries.

You are poetry books and quotation collections and brittle ironic guides from the sixties on how to be manly, placed purposefully on top of a copy of *For Whom the Bell Tolls*. You're also the grandly artful graphic novels *Jimmy Corrigan: The Smartest Kid on Earth* by Chris Ware; the work of Nate Powell; and *Here* by Richard McGuire, which is brilliant and trippy enough when you're *not* high.

Your Eccentric Fiction section at the table's corner: Murakami; *Inherent Vice* by Thomas Pynchon; *Beautiful Bureaucrat* by Helen Phillips, where shit with the faceless bureaucrat who has halitosis gets *real* bizarre in the best way. Then that piece by Gertrude Stein that people assumed they were too dim to understand, but turns out Gertrude's lover was jealous of a woman named May and made her change every instance of "may" to "can," rendering it kind of gibberish. But is it gibberish? Maybe it actually does make sense. Like, even *more* sense.

The room is filled with smoke and conversation. Everyone surrounds you, gazes at you, carefully choosing which of your books to pick up. (Also, from down here, it sounds like maybe they consistently drag your table to one side of the room, pick you up, and drop you and laugh.) Sometimes, your collection inspires an open mic night of sorts, each person taking turns standing up to read their favorite poem, Dorothy Parker quote, or short fiction by Raymond Carver. Other times, you prompt philosophical arguments of the polite but livid variety so familiar to book clubs and Facebook comment threads. Often, the discussions become so heated that Leto has to come out from the kitchen, where he's been cooking Stoner's Delight for the group and contemplating if it's possible to enjoy Hemingway as a feminist, to break it up.

I don't know. Maybe this is all my imagination, or the result of my bedroom being hot-boxed right now. Maybe you don't even exist. You might be a single *Maxim* on the floor next to the toilet. One tattered Hunter S. Thompson. A creepy, extensive collection of *Alice's Adventures in Wonderland* editions. Those *Magic Eye* books that you see different pictures in when you look at them with dead eyes. You could be anything. I will likely never know because I mean-mugged *Almost Famous* Coat once when she blocked my car in, thus branding myself Bitchy Cat Lady and severing my chances of being invited to the ongoing party.

But I love the you that could be. Even if you're just a dirty magazine and a Dan Brown paperback nobody's claiming; with time and luck, you'll grow. And if you mysteriously acquire a copy of *The Tao of the Dude,* you'll know who it's from.

Keep on Keepin' On (but also, could you keep it down?), Bitchy Cat Lady

**SOCIAL PROBLEMS--POLICE--
James, Stuart H., Paul E.
Kish, and T. Paulette Sutton**
--For Life, Scarred
--Recovering, Never

Dear *Principles of Bloodstain Pattern Analysis,*

Get the fuck out of here. Bwa-boogity-boogity (that's me shivering with disgust).

I don't know who put you on hold, or why. If it's the guy who offered to paint my likeness on the back of his bomber jacket, I'm going to flip my shit. Actually, if it's anyone but Gary Sinise from CSI or someone else with the trimmings of a qualified detective (trench coat, tiny pocket notepad, evil noir dame hanging limply from arm), it's gonna be super awkward. I'm not making eye contact.

Shit, what if it's some poor old lady who's mistaken you for a new Nora Roberts series? You could seriously give an unsuspecting person a heart attack. I know what you are and you still gave me what my nephew likes to call the Hot Burps. Bwug.

We need to get you out of here. None of the staff can stop reading you. We keep yelling "Oh my God!" and it's a library and we're supposed to be quiet. We're all in a circle around you, holding our hands up in the air every time someone turns the page, looking away with grimaces on our faces. It looks like a religious testimony in here. Or like Adele is singing to us.

Don't make us look anymore! Especially the head lacerations. ESPECIALLY all the blood-spattered toilets. You just gotta go. Blurk. Bwa.

You can come back when whoever put you on hold is done with you and we will gladly send you back to the library you came from. Just don't return with anything from your insides on your outsides. You know what I'm saying.

Bwuughck,
Annie

Dear *An Education,*

It is with great sorrow and a bit of trepidation, actually, that I feel I must send you away.

I'm trying to sound British, how'm I doin'?

You and I have a lot in common. Youthful zest and an eagerness to be unleashed into the big bustling world; ill-fated romances with untrustworthy men; and we both use the term "tremendous fun" often. Because, let's be real, we *are* tremendous fun.

It's because I see so much of myself in you that I have to give you up — to someone else I see myself in: my darling, innocent, eighteen-year-old niece. She needs a no-

nonsense talk about the nonsense of early womanhood. My hope is that you'll confess to her stories of your tryst with Slimy Older Man and she will find it so frankly described that she will not feel the need to repeat it. We're in the Midwest, you see, not London, so while your older beau introduced your heroine to highfalutin restaurants and posh hotels, the equivalent in our area is, like, a Coney Island restaurant and a Holiday Inn with a pool. I would prefer if her education came from you and not someone who won a free night's stay in a Jacuzzi room at the casino.

Tell her what you said about the importance of school, especially the stuff about learning to be kind and good and meeting kind and good men. But, you know, the actual learning part too. Your elegance and wit are just what this young woman needs to begin her new independent life. That and some mace. And a copy of *Reality Bites.* And laundry quarters. And some Taco Bell gift cards.

I trust you. Now go forth and provide, *An Education.*

Cheers,
Annie

DIARIES--World War II — Frank, Anne
--Girls, Brave

Dear *Anne Frank: The Diary of a Young Girl,*

Your author wrote down a wish to "go on living even after my death," and, through her tragedy, she got it. So maybe that means your wishes will be heard if you write them down. Here are mine for her:

That she could have seen her name in print.

That she could have had a long stretch.

That she could have hugged her mother tighter, and that she never had a moment's guilt about whatever their last words were.

That she could have marched with the feminists.

That she could have had grandchildren who

rolled their eyes at each other when she spoke of the war.

That she could have fallen in love with a handsome man who one day plucked a whisker from her chin, to which she could have responded, "It's hell getting old."

And I wish they could have had a good laugh about that.

With Appreciation and Love,
Annie

JUVENILE FICTION--
Silverstein, Shel
--Apples, Bitter

Dear *The Giving Tree,*

You broke my heart. It started out so sweet with you. I thought you were about being, you know, giving, and how generosity fills us all up with happiness. It was all little children skipping around in the woods — for a while.

In high school even, I loved you so much I gave you to a boy I also loved. Do you want to guess how that went, *Giving Tree*? Want to guess who was the tired old stump at the end of *that* book?

We didn't cross paths again until someone gave you to me for my baby shower. I arranged you on the shelf without much thought and didn't get around to rereading you until I found myself awake at 2 A.M. with another human literally sucking the life out of me. I've got to say, the glow of the crib-side lamp does not cast you in the most favorable light.

This tree you talk about keeps giving and giving and GIVING and you say she's happy, but I don't know. The little boy brands her with his initials, takes everything she has, and leaves. And she's happy about it? The End?

Maybe it was just the sleepless nights or the hormones, but WTF?

Tree was not happy. My girl was suffering. First off, she is the only tree in that goddamn forest. No other trees to commiserate with or advise her against all that giving. Then that punk comes and takes all her leaves and apples. In the end, I think we're dealing with a severe case of low self-esteem and an unreliable narrator.

I'm leaving this relationship with unanswered questions. Is the tree supposed to be a mother figure who enjoys sitting alone and waiting for people to ask her for stuff? Or is she, like, the Environment? Were you written so that future politicians could use you to illustrate points about using up natural resources?

I don't care. You took from me, *Giving*

Tree. You took and you took and you took. And I'm still giving, but now I'm giving you up.

How Do You Like Them Apples?
Annie

FICTION--Nin, Anaïs
--Psst
--Dirty Stuff

Dear *Delta of Venus,*

Two words: Summer. Thrillz. Beach sex without the sand burns. I'm talking about you, Anaïs Nin erotica.

I don't think they ever explained the difference between romance and erotica in library school. But now I know! Romance is: read thirty chapters of bickering and almost-kisses, and then watch the main characters do it. The end. Erotica is: pick any page — filthiest thing you've ever read.

You, *Venus,* somehow manage to be the dirtiest and the classiest of them all, and that's why, this summer, you're my shoreside secret. Just a plain beige paperback. No one could ever guess the astrologer exhibitionists, nude models, and masked woodland strangers that are getting it on in between your pages.

Here at the beach, I look like the unassuming reader in the one-piece, cloaked in multiple layers of SPF 70 and a big brimmed hat. A volleyball guy who jogged over to retrieve his ball just told me to smile and look like I'm enjoying myself. What a fool. I said nothing, just glared and violently snapped you back open in front of my face. But in my head I was telling him, "Yah, have fun hoping someone's top falls off during a serve. I'm reading about an orgy in an opium den. And I'll never wrinkle. Hate me cause you ain't me."

Anyway, thanks for your discretion. And for always referring to female genitalia as "her sex." That's some classy shit.

Enthusiastically,
Annie

HOBBIES--Allen, Joseph
--Appalachian Weaving
--Squandered Lives

Dear *The Leisure Alternatives Catalog, 1979,*

I dig you. I enjoy thinking about a time when being leisurely wasn't "Nnnhhhnnh. Must. Binge on. Netflix. Must. Read. One more article telling me why I'm a terrible parent."

You're an eccentric and seriously thorough book. You've thought of everything! Sports, nature, enlightenment, the arts! And you have wacky headings like "Give Yourself the Business" and "Canadian Sunburn" that are so intriguing. I really think a weaving class in Appalachia sounds chock-full of charm!

I would love to spend some downtime making kaleidoscopes and learning Polynesian dance. Or canning pickles and traveling in search of the Loch Ness monster. But the truth is, while writing this note, I've been interrupted five times to help my son load the couch cushions into his dump truck. I've answered two e-mails and read a list of things every '90s girl remembers. I could look into you later tonight, but I'm probably going to watch an episode of *Roseanne* that I've already seen thirty times.

Now I remember why I don't pick you up very often. It goes downhill fast with you, and I'm not talking about your suggestion to join an all-girl ski club.

Just looking at you makes me feel like I'm squandering my life. You remind me of those people I run into on Saturday afternoons who talk about all the shit they've already gotten done that day. And when I say that, I mean people that post their to-do lists on Facebook. That's "running into" now. I know it's pathetic, but what's to be done? We can't all be art-cinema buffs and sailing experts like you. God, if you had an Instagram, I'd have blocked your ass a long time ago.

There's really nothing left to say here. You keep doing you, and I'll do the me that doesn't fully live up to my potential, and we'll meet up next time I convince myself I want to start a consciousness-raising group.

Which is, give or take, every six years.

Bye for Now,
Annie

CARICATURES AND CARTOONS--
COMICS--Clowes, Daniel
--First Times

Dear *Ghost World,*

I don't know a more couth-ly way to say
this. You popped my graphic novel cherry.

I was curious, but I thought comics and
graphic novels were all about burly super-
heroes and epic battles. I like my reads
evenly paced with lots of character explora-
tion and subtext. The only thing I could
think of that's worse than having to watch a
bunch of action scenes is having to read
them.

And then you came around, a book with
goddamn GIRLS talking about perfecting
their look, and Don Knotts–lookalike as-
trologists, and which boys they secretly like.

You weren't a saccharine teen romance
and you weren't wordy prose, expanding

my mind and vocabulary. You were something different than what I'd been reading. You were relatable. I didn't know unique and smart and snarky and anxious girls were important enough to have a book about them. I wasn't even thinking of myself as an audience deserving of being reflected. They say your first changes you forever. And you did. You most certainly did.

Once I started, I couldn't stop. I wanted more books that felt like me. More talking like people actually talk and illustrations of awkward moments. More mundane suburbia and the crazy unmundane shit that happens in it and more of reading and seeing the realest emotion on the expressions of characters' faces.

So thanks for being my gateway drug.* I'll always think of you awash in the nostalgic dreamy teal of all my '90s memories.

Stay Sweet,
Annie

* For a list of books to read after *Ghost World*, see page 310.

YOUNG ADULT FICTION--Meyer, Stephenie
--Me, Bite

Dear Twilight Series,

You're on a card table at a yard sale, sitting next to an old Baby Alive doll and a plastic makeup organizer that says MADISON in foam sticky letters. Someone's not tagging along to college after all, hmm? How the mighty have fallen.

A little lonely now? Well it serves you right for hogging young Madison's prime leisure-reading years, rehashing the same old tropes, and never introducing her to something new.

Oh, he loves me! But he wants to kill me! But he really hates that about himself . . . But he loves me so much he'd rather die than be without me! But in order to keep him around I have to promise my mortal life to a vampire coven. Oh, look, we're pregnant! This birth is going to break most of my bones and my baby daughter is going to start dating my werewolf ex-boyfriend. Well, we do crazy things for love!

Gah, please don't figure out a way to tag along with her to campus. We can't be sending Madison off on her own with that kind of blueprint for romance.*

No, I'm not buying you. Stop looking at me.

Get a Life,
Annie

* But for a list of books about girls and romance that don't make me cringe, see page 310.

ENGLISH--STANDARD--Morehead, Philip D.
--Equivalents, Equivocal

Dear *Penguin Roget's College Thesaurus in Dictionary Form,*

I know you think I only use you when I need something. You're hurt because I don't include you in any of my favorite book lists, even though I turn to you more often than any of those other volumes, tomes, folios, pocket editions, and opuses.

But you are more than just a plunder call to me (eesh, I guess we're not on the same page with what "booty" means). You are so special. How do I say this without sounding banal? You're exceptional! You're singular! You are indispensable. See "necessity"! That's what you are to me. So file this letter under "love."

I've had you since I was a freshman in college or, as you might say, a tenderfoot in finishing school. You stayed by my side through the terrible fiction I wrote for my first writing class, when I thought plot meant every story had to include a car crash or an abortion. You helped me redefine public service when I was in library school: communal ministration or general duty. And upon graduation, you assisted me in amassing words for my résumé that made me seem more qualified for jobs without lying. I've never typed the word "competent" so much in my life! Those were crazy times.

You've been around for my entire adulthood, which means we have all the same references. For example (I know you love examples!), you are new enough to include "bunghole" but not modern enough to have entries for "cray" or "on fleek." Because who cares what the kids are saying these days? You're 782 pages — I'm pretty sure we can find a way to tell someone their eyebrows look nice. Like, "The mane on the summit of your head is winning." There. Easy.

It may be true that you are often overlooked. I feel like people assume you're boring. But they haven't spent hours reading your "color" entry like I have; cobalt, fir green, doeskin, ocher, aubergine. It makes me want to write a dystopian novel about Rainbow Brite. Your "Food" section compels me to depart from my desk, venture to the local java boutique, gorge myself on almond horns and apple pandowdies, and describe every bite to passersby as, "Delectable, my kindred spirits. In fact, decadently ambrosial."

That's why I get so handsy with you. I get excited when we're together. I get lost. In the best way. And maybe I don't show you off to my friends or recommend you at the library. But you have to believe me, I bring you with me everywhere, because you are always inside my head. When I want to tell someone that a book is "great" but stop and think of a better word. When I tell my toddler to say "Oh, man" instead of "Jesus Christ" (I swear I don't know where he got that). When I tell my husband that I'm not mad, I'm just mightily irritated. There's a difference. Look it up.

With Intimacy, Deep Affection,
and Ardor,
Annie

MEMOIR--ANIMAL--Grogan, John
--Dog Lovers, Suspicious Of

Dear *Marley and Me,*

Stay calm. I'm passing you this note to ask if you're being held against your will. This guy you're with, I'm not sure he's reading you for the right reasons. In fact, I would go so far as to say he's not reading you at all.

May I mention, and I hope you don't mind my saying so, that you have a pretty niche, if also large, audience, and that is people who love their dogs and refer to them as their fur children. In a Venn diagram situation, you would be inside that middle part labeled "People Who Love Their Fur Babies" and also "Readers Who Want to Feel Good and Be Done With It." And while it's possible that your reader falls into the latter category, I feel that a true

Fur Parent wouldn't leave their Fur Baby locked up at home in order to go to a bar and read about loving Fur Babies. Especially if their Fur Baby is of the crazy, drooling, bounding, untrained variety, like Marley. You're the kind of book that begs to be read at home, while petting a Fur Baby. See how much I'm saying Fur Baby? It's kind of annoying, isn't it? You might pass that tidbit onto your readers in the future.

This dude holding you has no dog hair on his button-up, going-out shirt. His shoes don't look gnawed upon. It started raining, and he didn't get up and pay his tab, explaining that his dog, like Marley, is afraid of thunderstorms. In fact, it's clear he's in it for the long haul here tonight. I know, because I've been staring at him for the past fifteen minutes and, though I am obviously unnerving him, he remains on his stool.

Okay, maybe you don't think that's enough to go on. Maybe he doesn't own a dog, but he told you he was reading you for book club or because his grandma gave you to him, which is a whole different *Marley and Me* Audience Venn Diagram I don't have time to get into.

Maybe this guy's just waiting for the bus and he happened to be carrying you with him. But if he's *really* reading you, then why

178

does he have his eyebrows furrowed? Like he's trying to digest something he's never thought of before? You're not Proust. You're about a dog that wrecks shit all the time. I've checked you out to sixth graders for book reports.

Also, why isn't he crying? I hated reading you (sorry). I spent the majority of my time with you rolling my eyes. But I still have a soul, so I cried when Jenny lost the baby and Marley consoled her. I still had to get the tissues at the end when it's, like, you know, The End. For Marley.

You are designed to make people cry, even people that don't like you. You're like that old Folgers commercial they play at Christmas when Peter comes home from college and makes coffee for his mom. She smells the coffee and cries out, "Peter! Oh, you're home!" And then any normal person begins crying and attempts to reconcile the anger they feel at an advertisement jerkin' their tears with the overwhelming happiness they felt when Peter came home. But this guy isn't weeping even a little bit. His only facial expressions are Confused by Book, Confused Checking Phone, Looking Up to See If She's Still Watching Me.

So, to sum up, the man you're with is either (*a*) neglecting his dog at home, (*b*)

Completely Dead Inside, or (*c*) using you to pick up chicks.

Or, it's possible, he's a decent guy with social anxiety who read a totally different book that told him to go out in public with a lovable memoir to help him start a conversation and make one friend. Now I've ogled him and probably scared him back into hermit life forever. I feel bad. Ach! *Marley and Me,* some of your feel-goodness has seeped in. I should go talk to him. I hate to admit it, *Marley,* but you've taught me a valuable lesson.

Lots of Love,
Annie

Dear *Marley and Me,*

Nope. Definitely (*c*).

I Knew It,
Annie

SUCCESS--PSYCHOLOGICAL ASPECTS--Covey, Stephen R.
--New Tricks, Old Dog

Dear *The Seven Habits of Highly Effective People,*

January 2005

I am so pumped about this! New Year, New Me, now let's get our *Seven Habits* oooonnnnnn like Donkey Kong. Okay, Part 1. That's just the "explaining why" part. Pass. Part 2. Cool graph, okay, got it. Here we go! Habit 1: Be Proactive. That's me!

You're kind of long, fella, aren't you? Lot of circle charts, too. I think, basically, you're just saying: Be proactive. Be committed. Make circles. Thirty-day test??? I don't have time for that. I'm going to skip ahead to habit 2.

"Habit 2: Begin with the End in Mind."

Pssh, got that right.

This is making me sleepy.

January 2008

I totally forgot I had you! I was doing really good working through your steps for a while. I remember there was, like, "Have a goal" and be a circle. Where was I? I had to be on part 3 already. "Paradigms of Interdependence": well, well. Fancy talk.

Oh, I like this: making deposits in an Emotional Bank Account. Six steps. Six steps *within* the seven habits? More with the making commitments? You're so naggy. Sorry, that was a withdrawal from your emotional bank.

January 2011

"Habit 6: Synergize." Synergize? That sounds sooooo boring.

I forgot how many fucking charts this thing has.

January 2014

Let's just start fresh. I don't remember where we last left off. Cool graph!

Whoops, here I go dozing again!

August 2016

Annie to Visiting Friend:

"Yeah, it's super insightful. I totally read it. Borrow it? Keep it! I practically know it by heart."

Effectively Yours,
Annie

DRAMA--20TH CENTURY--Miller, Arthur

--Drama, High School

Dear *The Crucible,*

A teenage patron just asked for you, and even though I haven't thought of you since you were required reading in my high school English class, I wonder how you are these days.

I read you grudgingly and was happy to leave you at the end of the term. First, you are a play we had to read aloud in our own halting monotone voices, which was terribly boring. But more than that, your story reflected back to me the bullshit of high school. You were an echo of all the bummers of being an adolescent: the gossip and judgment and hysteria, your punishments of death more violent, but no less dramatic, than the parking lot fights, the "accidental" shoulder shove in the hallway, and the threat of a bad reputation.

It occurs to me now that there's a reason our English teacher assigned you at exactly this point in our lives: to teach us a lesson about how words can take on a power of their own once they leave your mouth — as young Abigail's did when her accusation about her lover's wife backfired and resulted in his hanging. Or maybe it was a lesson about how judging others tears at the fabric of your community, and then everyone is pointing fingers at everyone else, and all your cows get loose in the mayhem. I gotta be honest, though. Even in the Advanced Placement class, such personal lessons were lost on us. It took us well into the third act to stop giggling every time someone said "Tituba."

Still, as I take you from the shelf now, I'm met with memories not of the dull readings and the dangling promise of a TV cart being rolled into the classroom to watch the movie version. Instead, I have nostalgia for you and I have to ask — did I give up on you too quickly?

For all the repetitive stage directions and long asides, there were afternoons when my crush might get stuck reading John Proctor's part, and I could daydream about him speaking the words "I may think of you softly from time to time" to me. Also, my friend Jack and I got a lot of mileage out of passing notes that accused each other of being in the woods with the devil. We called everyone Goody So and So for the rest of the semester. So you couldn't have been all that bad. We had some good times, right?

Since I saw you last, I've read more about the Salem witch trials in Stacy Schiff's *The Witches: Salem, 1692.* You two should get together. I learned that the Puritans were hitting the cider pretty hard back then and had gone through a harsh winter. Indeed, they were probably suffering from seasonal affective disorder, which, I know from experience, can make you a delusional nutjob. That information might have punched up our drowsy performance of you. The courtroom scene where Abigail accuses Mary of shape-shifting into a bird may have been particularly entertaining with slurred speech.

Do you think we could give it another shot, now that I'm a little older and not too distracted by teenage drama to absorb your tale of fanatical faith and deceit? Are there more like you I've cast aside and forgotten because of my hasty heart and infatuation with MTV? Should I dig out *The Odyssey*?

Am I ready for Virginia Woolf? I'll never know unless I try again.

For now, I have to send you away with my patron, this less-than-thrilled sophomore boy, but I promise when you get back, we'll give 'er a go.

See You (Way) After Class,
Annie

JUVENILE SECTION
--Jobs, Very Important

Dear Public Library Children's Section,

You make it look easy, like fun even. But what you do is hard work. Important work. And you're the only one that can do it.

Kids come to you for lots of different reasons. Because they need a biography for a book report, a superhero comic, or the next *Dork Diaries.* Because it's too hot outside. Because they can be loud. Because they need quiet. Because their parents dragged them. Because no one is taking care of them. Because you're a joyful space. Because you're a safe space.

Hard work. These kids have got to fall in love with you. They need to learn to read, so they can love to read, so they can understand how many different lives they are capable of. It's VERY important. More than the Philosophy section and the College Test Prep books and the Nobel Prize–winning novels upstairs. Because no one's going to be there to read or write those other books if you don't pull your weight.

So show off your Maurice Sendak, your Jacqueline Woodson, and Lois Lowry. The kids will see themselves in you. *Amazing Grace* and *Stuart Little* will tell them how brave they can be. Your nonfiction section can provide advice without judging. And your chapter books can be counted on to deliver the whimsy. Be at the ready from morning to night and on weekends too. Be a place of peaceful comfort and rowdy imagination and encourage lots of plan making for the future. Don't ask anything in return. You have to give it all away.

You're not getting much help from the rest of the world. I don't know if you've been talking with the Current Event books, but it's pretty disheartening out there. So you've really got to rally. Stand up straighter on the shelves. Try to make your titles visible through all that book tape. Hold it together.

Take Care (like really good care),

Annie

Dear *Pictorial Anatomy of the Cat,*

I don't even know how you got here. Without your book jacket on — which is who knows where — one might have assumed you were some sort of mythical fairy tale about kitties. But in reality your insides . . . are about insides.

Don't get me wrong. I thought your lateral view of the abdominal viscera was neat. But, can I say something? You're creeping people out. You go on and on about the cutaneous maximus. This is a public library. No one here knows what that shit means.

Where will you go from here? It's difficult to say. We could put you in the annual book sale, but we know the one guy who would buy a book about cat dissection, and he's been permanently banned. Your various depictions of gross cat claws with the skin ripped away might make for good Halloween cards but, truth be told, I don't give cards out on Halloween. And I just won't go there for Valentine's Day.

So . . .

Go 'Way Now,
Annie

SHORT STORIES--Brautigan, Richard
--Shtick, Librarian

Dear *Revenge of the Lawn,*

You're a good book. You are. But you gotta go. Don't ask why.

Okay, here's why: I got you from a bartender in Chicago whom I was probably going to make out with, but then I started dating my husband and ditched the bartender and never gave you back. I remember it was in October because Michael texted me to ask what I was doing that weekend and I told him I was going to carve pumpkins and maybe break a date with this guy, your original owner. And he replied, "Have fun gutting your pumpkin and your bartender." And I thought it was rather clever and flirty and, well, that was the beginning of the end for you. In my fickle heart, you were already on your way back to where you came from.

Where you came from, I imagine, is behind a bar in a stack of other yellowed literary paperbacks just like you, waiting for the next Pretty Young Thing to order a dirty martini and mention that she likes to read. The librarian shtick really works fellas into a lather. And I did have a little bit of a Parker Posey in *Party Girl* thing going on back then. Plus I hung out at that bar a *lot*. It was only a matter of time before your quirky satirical stories fell into my delicate bookish hands.

But the thing is, that was, like, almost a decade ago. I'm married now. I've got a kid. I don't know how much almost-old-boyfriend stuff I should have lying around. The other day my son pulled you off of the shelf and gnawed on your corners and, to be honest, I don't know where you've been.

We had some laughs, but, in the end, we don't belong together. Let's just leave it at that. I'm sending you back to the address written in pen on your endpapers — some guy named Mel in Austin — and I'm not looking back.

Don't Call Me,
Annie

**ATHLETICS AND SPORTS--WINTER
SPORTS--ICE SKATING--
Montgomery, Richard
--Sweeney, D. B.**

Dear *The Ice-Skater's Bible,* 1982 edition,
  The only reason I'm keeping you is out of
a fierce loyalty to the award-winning film
The Cutting Edge.★

Tooooeeee Pick,
Annie

★ Award received: Olympic Gold, pairs program.

CULTURE AND INSTITUTIONS-- SEXUALITY--Keller, Wendy
--Mind, Open
--Legs, Closed

Dear *Cult of the Born-Again Virgin: How Single Women Can Reclaim Their Sexual Power,*

Don't take this the wrong way. You're going to take this the wrong way, aren't you? We took you out of circulation at the library.

This isn't about the women dressed in togas, staring out at the ocean on your cover. And it's not that I disagree with reclaiming your sexual power or the whole "Keep Your Knees Together — Change Your Life" philosophy. I mean, I *totally* disagree with you that depriving yourself of sex when you like sex will help you land a better man or "serve humanity" or "restore a sense of dignity and elegance" to your life. But I'm a librarian and I don't base our collection on my own opinions. There are other reasons for your dismissal.

The thing is, you're bringin' a lot of baggage. Some of your names and locations have been changed to ensure anonymity, but your dedication page gave the first *and* last name of the author's ex-husband (who was thanked for breaking said author's heart). You seem just the tiniest slivery-est bit — bitter. And your advice on how not to dress slutty comes off a little harsh. Plus, it's hard for me to pull off a Princess Diana or Helen Hunt vibe; and you don't give any other examples of nonslutty women.

Let's focus on the positive here. You offer some delightful quizzes. Before we discarded you, someone who had checked you out wrote in you, and I got a big tickle reading their quiz responses. In the passage where you ask the reader to "check off traits you secretly like about yourself,"* this reader noted that she has a strong immune system, buys only quality items, and loves her exercise program. Sadly, there was no check

* Of course, if you like something about yourself, it has to be "secretly." Why would women openly like things about themselves? That's ridiculous.

mark next to "I am logical."

In the "Love Style" quiz, pencil marks indicate that this closed-knee wannabe actually managed equally high scores for Narcissistic Lover and Self-Sacrificing Lover:

1. "I like to keep my partner just a little bit insecure about how much I care." **True**
2. "I recognize that to be truly loving often means sacrificing my wishes and goals." **True**
3. "I am friends with almost all my old lovers." **False**

(I scored Codependent Lover, BTW. Because I'm friends with some of my old boyfriends and like my lover to keep in touch throughout the day.)

I hope it doesn't make you angry that I brought you home. You haven't been home with anyone in a looooooong time. Sorry! Inappropriate. I save you (just like you're saving yourself — gah! I'll stop, I promise) for this thing I do every year where I have my girlfriends over to do crafts, but we mostly end up eating snacks and drinking too much. I think you would love it. I do a dramatic reading of some of your passages and we all do your quizzes. Sometimes, we even drink enough to scare up some togas. Some girl talk is just what you need to "Stop the Insanity," which is a Susan Powter quote from your book.

Keep in Touch (if you know whatamean),

Annie

JUVENILE FICTION--Howe, James
--Vampire Bunnies, Used

Dear *Bunnicula,*

Like everything else that the youngest of five children receives, you were a hand-me-down. Two of my three older sisters read you, then my big brother. Eventually, one of them, probably when bribed with the promise of a brand-new book from the Scholastic catalog (which I NEVER fucking got), threw you at my face while they whizzed by on their new bike in their new Kmart jeans, and I was grateful.

I took such pride in you that I marked you with my "Viola Swamp Is Watching You" stamp on your inside cover. And, as you can tell from the blobby schmear on your title page, I slammed you shut before the ink even had time to dry. Because you scared the bejesus out of me.

First, you're about a vampire bunny. Second, your other main characters are a dog and a cat, and I was terrified of both animals when I was a child (in addition to having no earthly possessions, I was attacked by a tabby as a toddler and nearly sacrificed my eye — I was basically feral). So your cover was off-putting, to say the least, with terrifying pets and that vampire bunny's red eyes and fangs. Then the editor's note claiming that you were a true story? It was too much for my tender soul. I shoved you under my bed, an offering to the alligators and at-large *Unsolved Mysteries* suspects I knew for a fact to be living there, and went back to my *Sideways Stories from Wayside School.*

It wasn't until I was an adult that I found you again at my parents' house and realized I had finally gained the courage to open you back up. It took me twenty-two years, *Bunnicula,* but I am ready now. I'm ready for us.

Of course, if anyone had told me, which of course they wouldn't have because they were too busy going to concerts I wasn't allowed at and buying candy necklaces that only came in four-packs, that the only thing your vampire bunny sucks is juice from fruits and vegetables, NOT BLOOD FROM

YOUNG GIRLS, I could have given you a whirl when I was at the appropriate reading level. I'm sorry it took me so long to come around. But I'm here now. And that counts for something, right? Let's do this. Your cat character is declawed right?

Love Your Final Owner,
Annie

FICTION--Gilman, Charlotte Perkins
--Room, Trapped In
--Small Talk, Better Than

Dear *The Yellow Wallpaper,*

Friday afternoon
I get the kitsch of putting you in an upstairs B and B room with yellow wallpaper, but, geez, I wonder if your owners know what you're about, or if they think you're an antique decorating book. You don't scream "vacation read." You scream "Lemme out!" Ha!

Friday evening
Apparently, people that stay at bed-and-breakfasts go to bed at eight o'clock. Michael doesn't want to make noise and wake Lois and Don in the next room. I don't

want to leave and risk common-area small talk. So we're going to read you to each other.

Later Friday evening
Welp, turns out reading you doesn't make me feel better about being in a confined area. Guess I better turn in. Rest always cures what ails you, right *Yellow?* Good old rest cure.

Saturday morning
Breakfast was served. Turns out we both hate strata. And it's raining. This place is a bust, but we paid for the whole weekend. Does it smell funny in here to you? I'm gonna take a nap.

Saturday afternoon
Didn't I leave you on the nightstand? What are you doing over here by the door? Are you trying to tell me something? The husband says it's my imagination, which I say sounds a *bit* familiar, don't you think?

Saturday evening
Cocktail hour in the common area. Liberate me now!

Sunday morning

Michael came in and found me crawling around on the floor. I was looking for my earring, I swear! Anyway, we're blowing this Popsicle stand and you're coming with. I'm replacing you with an Anita Shreve novel and getting the hell out of here.

Let's Boogie,
Annie

HEALTH AND HYGIENE--Food--
Excess Of
--Self Respect, Get Some

Dear *Better Homes and Gardens Dieting for One,*

Though you sit among my Martha Stewarts and kitschy old cookbooks now, I found you in the free-book bin. Because the relative of whoever died and left you in their house looked at you and thought, "Here's another cookbook from the 1980s with disgusting food photography and an excessive addition of fruit to meat dishes."

I'm going to give it to you straight: that person was not wrong.

Your title is one strategic ellipsis away from a Katherine Heigl rom-com about a woman who falls in love with her produce man (Paul Rudd) on her quest to reinvent herself (aka lose weight), even though she was originally gunning for her Bikram yoga instructor (Channing Tatum), except we already know from the trailer that she farted in class and blew her chances — pun intended. "Thanks for the dietary fiber, Paul Rudd! Now I'll never be the hot yoga lady at my class reunion!" Spoiler: Paul thinks she's already beautiful and he's going to show her how versatile beets can be in a menu. Roll credits.

But it's not just your title. It's not just the shriveled silver sticker clinging to your cover ("$5.95! A hit of a gift!"). It's the whole way you present yourself. You devote a two-page spread to a photo of a lady in a floppy-bowed blouse answering her phone on her lunch break even though she's attempting to enjoy her skim-milk-and-shrimp pocket. Is that the face of *Dieting for One*? Probably. Yes. Even so, the newer diet books only show pictures of extremely trim people, and they're usually not shown eating because they are laughing so joyfully at their fitness that they can't get the fork in their mouths. That's what you're competing with.

Have some self-respect! You have a lot going for you. You may not be the first to call your collection of recipes "de-light-ful" (oh, *light,* I get it). But, damn, your first recipe is a STEAK dinner. Own that! Take pride in your cheese-spread section, "Spread the Good News." It's one of a kind! Today's diet books have abolished cheese and carbs. But you embrace them! So what if your Polynesian Ham Patty looks dis-gust-ing. Your Reuben Sizzler sounds amazeballs. You have excellent instructions on how to start a sprout garden, and you were touting whole foods in 1984 when everyone else was like, "Whatever, just slap some margarine on it. Margarine will save us!!!"

You keep telling your readers they deserve delicious food. Aren't you deserving as well? Of a chance to teach more folks about the low-calorie benefits of liver and how to stuff damn near everything into a homemade cream puff? Yes. I believe you are. But — no. No one else in my peer group does. They laughed you right out of the room. They liked my screenplay idea, though.

The proverbial fat lady has sung, and she's lamenting your spicy beef cups. I didn't think your self-esteem could take the blow of the Free Box. And I want to try your lemon cheesecake recipe (though to say I'm skeptical of a cake that uses cottage cheese would be an understatement). Looks like I'm the "one" in your *For One.* At your suggestion, I'll lay out a single place setting

with my best dishes and really take the time
to savor you.

Welcome Home!
Annie

Dear *Forever,*

I've just discovered you right where you belong. In a secret stash of old journals and love letters and summer camp name tags from boys who promised to write (I know you know how that feels).

I wonder, in the forty-plus years you've been around, how many secret places have you been? Shoved in back of a dresser drawer, a forewarning to the neighboring Days of the Week underwear. Tucked under a twin-bed mattress. Camouflaged behind homework in the pocket of a Lisa Frank binder. Nobody had the guts to ask their parents to buy you, yet somehow everyone has read you. Around the world, females from twelve to fifty still snicker to themselves when they hear the name Ralph. You are womanhood's worst kept secret — a passed note recited in front of the class.

When I first saw you in 1996, I had so many questions. Like, what's a fondue party? Why couldn't I have a little sister that embroidered my jeans for me? When are we going to get to the dirty par— Ho! There we go!

I don't remember who gave you to me — you were probably one of my older sisters' or I stole you from the senior citizen center I volunteered at, which is how I got most of my raunchy reads. I do remember that I read you in one sitting, cross-legged on the floor, completely consumed with Kath and Michael's romance. How thrilling that a boy would like Kath so much that he bought her a gold locket! He told her she was delicious when they kissed! Did my hair smell as good as Kath's? And how could I get it to smell better? And, after that, how could I get boys to smell it?

I was angry with Kath's parents — that flabby librarian mom (you *thrice* described her as such) and her old-fashioned dad — for not believing in their daughter's pledged love. They said "forever"! They signed every letter that way. Until the letters stopped. Until the parents' evil plan to separate Kath and Michael worked, and Tennis Camp Theo came into the picture. That was a god-damn plot twist to beat the band.

In terms of the horny stuff, you delivered! I took mental notes for my future: buy bikini underwear; plan possible weekend ski trips to Vail around your period; for reasons not specified, only do it on rugs; find out what the clap is. Then I officially joined the club of People That Have Read *Forever* and could feel free to continuously make inside jokes in front of people who hadn't read you yet. It was fabulous.

So looking at you now, I was hoping for a little bit of That Old Feeling. First love: so romantic. Except — it's gone. The feeling. I can't bring it up. And in its place, I just feel Old.

I now find Michael's character sooooooooo annoying! He throws snowballs at Kath, calls her a tease, and makes comments about her body every five seconds. He wears an aftershave called "Mustache," for Christ's sake, which I now know for a fact he totally *did* put on his genitals, no matter what he claimed. Half the time he can't get it up. The other half of the time he can't, you know, contain himself. In short, he is a perfectly accurate description of a seventeen-year-old boy. Now Theo, on the other hand, has an *actual* mustache, asks Kath how she's feeling, and dances close without crude comments. He's a grown-up. And, shit, now I'm one, too.

If you hadn't guessed already, *Forever,* I am now Flabby Librarian Mom. And upon closer reading: of *course* Kath's folks are going to send their newly sexually active daughter away from her twerpy boyfriend after he got drunk and barfed at their house on his eighteenth birthday! They're no fools. Kath, though, my girlhood role model, now sounds awfully righteous when she talks "forever" to them. How could someone who was once the pinnacle of maturity for me now sound so childish? You're tricky, *Forever.*

This is to say nothing of the fact that, in my first reading, I completely blocked out all of your side plots: Sybil gets teen pregnant? Artie attempts suicide? There's a detailed cervical exam and discussion of safe sex? News to me. If your pages didn't say "Ralph" on them, I've no memory of their content.

So kudos to you, *Forever,* for staying true to your audience. But I'm not one of them anymore. Will I ever be able to access those blissfully ignorant shivers of first crushes and jean jammin'? Is what the original

212

Flabby Librarian Mom said true? Can you never go back to holding hands?

To continue the tradition, I'm going to highlight your dirty parts in pink, and hide you in a place where a new teen girl can discover you and keep you like a secret. Then I'm going to go upstairs and ask my husband to smell my hair.

Not Forever,
(Which I think was
your whole point, right?)
Annie

FILM--Beauty and the Beast
--Imaginary Libraries
--Bookdiggers

Dear Magnificent Library Featured in *Beauty and the Beast* Movie,

Belle is my favorite Disney heroine. For starters, she's the only "princess" who spends the majority of the movie with her hair in a dowdy ponytail because she could care less. She's just trying to finish her book. In my opinion, the premise of the film is: people constantly interrupt Belle while she's tryna read.

She finally marries the Beast because she's a total bookdigger and wisely realizes that the castle is big enough for her to hide in one wing, reading in silence and ordering that sniveling Cogsworth to bring her snacks and tell Beast she's got period cramps and is not to be disturbed. At the end of the movie, she's like, "This twirling around shit can end anytime. You guys know where the door is? I've got a booklist to start."

But the main reason she's my favorite is you, Library. You're so golden and glorious, towering over everyone with your endless rows of books. To be Belle for a day! The closest I get is the opening shift at the public library, which, on the right morning, can vaguely resemble the first scene in the movie, where Belle is walking into her provincial town.

At first it's all sunshine and birds chirping. I've had my coffee and I'm thinking about making a fun "If You Like Junot Diaz . . ." bookmark. But then, like Belle, I keep getting stopped by charming townspeople. Except instead of singing about how pretty but alarmingly literate I am, they're yelling at me because Joel Osteen's *Become a Better You* is checked out, or asking me to describe how to write a cursive "q" over the phone, or threatening to strangle another charming townsperson with their backpack straps (main reasons for violent threats between patrons at the public library are: Someone's Been on the Computer Too Long, Political Arguments, Disagreements About Whether Pit Bulls Are Sweet Lil Critters or Face-Eating Monsters — in that order).

You actually might be part of the reason I took this job. Because of you, I thought spending time in libraries was all wheelie ladders and silent, sustained reading time. And though I've since discovered that library patrons can be "mean, coarse, and unrefined," just as often they are "dear and so unsure."

When it's tempting to roll my eyes or put the LIBRARIAN HAS STEPPED AWAY sign out and pretend to be making copies, but really just finish my chapter, instead I put on my best Disney princess smile and forge ahead because that's what Belle did. She suffered through Gaston putting his filthy boots on the book she was reading, and she hugged Beast even though he probably smelled like wet dog. In the end she was rewarded with you — and, yeah, true love and living in a castle and whatever.

That's the dream. I'll keep wishing on a star that someday my fantasy library will come.

Until Then,
Annie

MEMOIR--Smith, Patti
--Vicariously, Living

Dear *Just Kids,*

A quick note. I'm so smitten with you, I can't help but pick you up when I'm waiting for tea to boil or brushing my teeth. And especially in bed. Actually, in bed is where the trouble begins.

My husband is jealous of you. I can't blame him — I am too. You're a delicate portrayal of two artists, equally hungry for food and fame, coursing through the veins of 1970s NYC. Sometimes, your hero and heroine are on top of the world. Sometimes, they live in a dump. Always, they have style. Always, they love each other. I want to be there with them! But Michael wants me here. I've been reading you in all of my spare moments. I haven't heard a word he's said to me for two days.

Now the hubs and I have got a pizza-and-movie-in-bed date planned (that's bohemian, right?), and I need to put you down.

Patti's love for Robert is palpably tender, and her love for art as wide as the sky. Your description of their youth leaves me with a lingering daydreamer's feeling — probably not going to last long 'cause we're watching *Smokey and the Bandit.* Sigh. What we do for love, right, *Kids*?

Now isn't the time for us. We'll have to take our affair elsewhere. I'll meet you in the break room at work. I'll be the one in the corner holding, well, you.

Faithfully,
Annie

BUILDING--Plans and Specifications
--Lifelong Dream
--Snacks

Dear Future Book Collection,

I don't know when you'll materialize, but it will happen in one of two ways: either I will one day be able to repay my student loans and build you, just in time to die of old age, or some foolish and powerful character will succeed in stamping out literacy once and for all. In the latter case, I'll be able to Dumpster-dive for shelving and discarded "paper symbol doo-hickeys" to curate you. Then for centuries after I'm gone, people will visit you, but only as a corn-maze-type attraction.

For now, you'll have to exist in my mind alone. But I've got you figured out: a room big enough for all of you to fit but small enough to feel cozy and hidden from the rest of the world. Leather couches that are also comfortable, piled with ratty home-made blankets, and a table scattered with empty tea and coffee cups (because, honestly, even in my fantasies, I'm pretty gross). The soundproof walls of the room will be papered in old American Libraries "Read" posters — most prominently, the one with Paul Newman lying on a pool table.

A locked door will lead into the room. This will be essential. There will be a doorbell; however, entry is denied unless visitor-candidates answer a series of questions posed by an exasperated robot voice: "Did you just put something in your mouth?" "Did you check in the pants you wore last night?" "Can I just get a minute here for Christ's sake?" Basically, my top three phrases.

Once the riffraff is sorted through, serious visitors will be allowed inside if they agree to speak only about books or to remain silent and bring a cheesy snack. And, what the hell, here too is a *My So-Called Life*–era Jared Leto! But he doesn't feel like talking; he just wants to stare into space and let me play with his hair while I read. Every so often, when I'm taken aback by something brilliant I've read, like my favorite Annie

Dillard quote from *For the Time Being* — "There are 1,198,500,000 people alive now in China. To get a feel for what that means, simply take yourself — in all your singularity, importance, complexity, and love — and multiply by 1,198,500,000. See? Nothing to it" — I'll pore over the dreamy pools of distant blue in Leto's eyes while I mull it over.

I'll also have a Reading Room Wardrobe, full of clothing Anjelica Huston would wear, which is to say something dramatic and stylish that says "I'm not fucking around." Drapey shit. Lots of otherwise beautiful pieces marred with cigarette burns. Sinister hats. That kind of thing.

But, I mean, obviously, it'll be all about you — the books. You'll all be wrapped in thick pastel marbleized paper, with your titles and authors embossed in gold lettering on the spine. Some of you will also be covered in Mylar plastic because I find wrapping books in Mylar very soothing. You'll be placed on shelves and in artfully haphazard piles around the room. And because part of the majesty of a library is not knowing where anything is and being guided by a radiantly clever but aloof custodian in a caftan (that's me) while Edith Piaf blares in all directions (I'll play exclusively Edith Piaf and the Shangri-Las), there will be no Dewey Decimal System. Instead, you will be organized by my dream categorization: Emotion.

For an example of how this system works, let's start randomly at The Mean Reds: books to cure the sneaky, all-consuming fear aptly coined in Truman Capote's *Breakfast at Tiffany's.* On these shelves, I will place the books that soothe me: *RFK: Selected Speeches,* by Robert F. Kennedy; *Girl Talk: Unsolicited Advice for Modern Ladies,* by Christie Young; Buddhist stuff; and anything Nora Ephron wrote.

Next to these books is the section called Make That Change, providing me with the strength to go forth and conquer. This area also comes with a little mat: when you step on it, the choir from Michael Jackson's "Man in the Mirror" sings "Make that chaaaaange." Books like *The Giver* by Lois Lowry, *The War of Art* by Steven Pressfield, and *The Gigantic Beard That Was Evil* by Stephen Collins will sit side by side.

Conversely, if I don't feel like making that change, there will be a Fuck It, I'm Just Going to Read Instead of Do What I Need to Do Today section: *Little Women* by Louisa May Alcott, and *The Dud Avocado* by Elaine Dundy. And a Fuck It (General) section:

the Ripley series by Patricia Highsmith, some Douglas Coupland, all of David Sedaris. This section butts up to (Worthwhile) Films Based on Books, so if I'm feeling particularly listless, I can just go limp on the couch and watch on repeat the scene from *Field of Dreams* where sassy Annie tells off the lady that wants to censor books. Somewhere in this section, I'll have to find a place for Watched *Game of Thrones* Before Bed Again and Can't Sleep (exclusively Maeve Binchy, to assuage the nerves).

There will be more positive sections, of course. I'm a sucker for Reminiscing About First Loves (*Eleanor and Park* by Rainbow Rowell) and Books That Make Me Feel Like I'm Hanging with a Good Girlfriend (*The Folded Clock* by Heidi Julavits and *No One Belongs Here More Than You* by Miranda July). Hanging with girlfriends will remind me of my sisters and the Comforts of Home section (*Mr. Shaw's Shipshape Shoeshop* by Eve Titus, Erma Bombeck, and encyclopedias). That will make me feel Midwesterny (Tom Drury and Marilynne Robinson).

There are the harder-to-pin-down areas as well. I know it's not an emotion, but my book kryptonite is the genre Crazy Surreal Shit in an Otherwise Mundane Setting: *Atmospheric Disturbances* by Rivka Galchen,

The Leftovers by Tom Perrotta, *Spilling Clarence* by Anne Ursu, *The Fever* by Megan Abbott. I love the plot of "one thing weird happens in a normal place" like some people love Scandinavian mysteries and books about what your face shape means (me, again). There's also Beautiful Freaks (lyrical and terrifying reads) like *Duplex* by Kathryn Davis, and my collection of novels about "Charismatic but Unstable Preacher Characters." There will be the Am I Crying Through Laughter or Laughing Through Tears? area: *Kiss Me Like a Stranger* by Gene Wilder, *Encyclopedia of an Ordinary Life* by Amy Krouse Rosenthal, most of John Irving. And that Happy-Sad Feeling you get when your characters don't end up where you want them to be but you're glad for the journey: *Bridges of Madison County* by Robert James Waller or, God, *Love in the Time of Cholera* by Gabriel García Márquez.

And that's not all. Those of you I've already read will have thin paper scrolls that roll out of you and remind me what you're about with sample book-discussion questions. I'll have traveled back in time and asked Grace Paley to sign all of my copies of her work and, while we're at it, maybe include some personal notes encouraging my writing and *maybe* even an unpublished poem dedicated to me. The drool stains on my Jeffrey Eugenides volumes will have magically disappeared. I'll have the same Virginia Woolf books I already own, but, by some unknown means, I will understand them (sorry).

Finally, in the back corner will be a small low shelf, admittedly dusty, unused but not unloved, with all my old flames. My disgusting *Pictorial Anatomy of the Cat,* my brokedown *The Goldfinch,* my disregarded *Cult of the Born-Again Virgin,* that book about trucks I hid from my child years ago that doesn't seem quite as annoying as I once thought. What this shelf has lost in face time with readers, it will have gained in nostalgia. "No two persons ever read the same book — Edmund Wilson," a note taped to the shelf will say, encouraging visitors to give these castaways a second chance.

Future Book Collection, like Diana Ross said but probably wasn't sincere about when she left the Supremes:

Someday, We'll Be Together,

Annie

■ ■ ■ ■

II.
SPECIAL SUBJECTS — LIBRARY EMPLOYEES — ASSISTANCE TO READERS

■ ■ ■ ■

"Now, I threw in *The Dual Voice Sings.* Just for fun. It's a book of songs composed exclusively by twins."

—Parker Posey as Mary
in *Party Girl,* 1995

I'd Rather Be Reading:
EXCUSES TO TELL YOUR FRIENDS SO YOU CAN STAY HOME WITH YOUR BOOKS

If you're reading this, chances are you've experienced bouts of antisocial behavior and "read rage" outbursts toward friends and family. Because, although you participate in society as much as necessary to convince your mom and the shrink your mom hired that you're not a shut-in, truth is, you'd rather be reading than doing just about anything. Did you think it was just you? It's not. You've just never met any of the others, because we don't want to talk to you either.

If you want to be left alone during the big clambake next weekend, here's a list of excuses to give and books that lend your lie a hint of truth.

"Got a hot date . . ."

. . . with a book. Reading can get you more hot and bothered than a Tinder date, without the cost of drinks and with a lower frequency of unwanted dick pics. I find

Dangerous Liaisons by Pierre Choderlos de Laclos a sultry book to hole up with. Its characters' devious acts of seduction will confirm that it's futile to leave the house in hopes of finding a mate and has been since 1782, when Laclos penned the novel. If you think the people you told about your imaginary date will want to know dirty details, steal some from Isaac Oliver's *Intimacy Idiot,* a sharp and truly funny book of essays about Oliver's young-gay-man-about-town escapades. Or get clinical with *Bonk* by Mary Roach, a scientific study of sexual intimacy that Roach spent two years investigating.

"Just hacked something up. Gonna inspect it."

If you're going with the sick excuse, it has to sound contagious enough that your friends won't want to come near you, but not so scary that they're going to want to check on you. I usually go with pink eye. Thankfully, choosing a book about disease isn't nearly as difficult as faking one. There are a bevy of mysterious-illness books. I prefer those of the sudden pandemic variety. Chances are you've read some of them without even meaning to: from *The Plague* by Albert Camus, to *Station Eleven* by Emily St. John Mandel. There is some really good writing within this minigenre, probably because writers are antisocial too and can easily imagine the rest of society as a giant infectious scourge. *The Dog Stars,* by Peter Heller, features a main character who is one of the few survivors of a widespread flu holed up in an airport with a crazy dude and a dog until he hears a radio message that gives him hope of life beyond his outpost. *Salvation City* by Sigrid Nunez tells the story of another superflu survivor. This time a young man survives in the small evangelical town he was taken into when his family died but still struggles with memories of his deceased loved ones and his dream for a future different from those of his rapture-awaiting friends and neighbors.

"I'm too drunk/stoned already. Can't leave the house."

This is my favorite excuse and, depending on your personality and past behavior, could be the most believable. Again, there is a fine line between good excuse and appalling deception for the substance-abuse lie. Veer more toward "Shit man, today's weed is stronger than that shwag we smoked in high school!" and away from "Oops, I swallowed a bottle of pills." Good drug books are as plentiful as medicinal marijuana clinics with terrible names, ranging from darkly hilarious to just plain dark. *Blackout: Remembering the Things I Drank to Forget* by Sarah Hepola is a memoir, startling in its humor and candor, about Hepola's path to sobriety, which includes a closer look at her drinking days. *Dopefiend* by Donald Goines is a tragic novel of a heroin junkie in 1970s Detroit and was written by Goines, a heroin addict himself, while doing time in prison. It is an honest and honestly terrifying book.

"Sorry — family obligations."

Keep it vague. But if they press you for details, borrow some from the following family sagas. Angela Flournoy's *The Turner House* takes place on Detroit's East Side, where a group of siblings gather to decide how to deal with the aging matriarch of the family and the home they were raised in, which remains unoccupied, in a neighborhood that has fallen on hard times. *The Lowland* by Jhumpa Lahiri features two devoted and dissimilar brothers in India in the 1960s. When one dies during a political uprising, the other tries to fill the void he left by marrying the deceased brother's pregnant wife. Okay, don't borrow *that* detail. You're probably not going to pull off

"I can't come to your birthday party because I have to marry my brother's wife and help raise their as-yet-unborn baby." But figure it out, because you're going to want to read these excellent books.

"I'll be at my high school reunion. Want to be my date?"

Asking if your pal wants to come along lends credibility to your fat lie and no one wants to go sit at a table with a bunch of strangers talking about their kids and their old grudges. You'll be safe from them accepting the offer and free to stay home and enjoy *Home Land* by Sam Lipsyte, a novel about a guy whose high school nickname was Teabag. That's reason enough to read it. But if you need more, Teabag begins writing in to his alumni newsletter with an unexpected honesty about what he's really up to, which is nothing to write home about. Lipsyte is a sharp and witty writer, and this is a book to make you snort. For a nonfiction reunion narrative, try *Bullies: A Friendship* by Alex Abramovich, who moved to Oakland, California, in order to live in the dangerous, intoxicated world of his former bully, Trevor, and forges an unlikely friendship with his high school tormenter.

"Alas, I've nothing to wear."

Say it just like that too. Then hang up or, if you're fibbing in person, spin around dramatically and walk away. You can experience the thrill of awesome fashion in your sweatpants by picking up *Vintage Black Glamour,* Nichelle Gainer's goddamn gorgeous book of photographs of black artists, both well known and overlooked, throughout the early twentieth century. Looks range from street style to disco to glamour pin-up. For a behind-the-scenes look into fashion history and a walk down '90s memory lane, read *Champagne Supernovas: Kate Moss, Marc Jacobs, Alexander McQueen and the '90s Renegades Who Remade Fashion* by Maureen Callahan. It's glitzy, it's dramatic, it's gossipy, it probably beats whatever plans you bailed on in order to stay home and read it.

"I have the menstrual cramps."

Works on everyone. Women will have sympathy for you. Men will immediately halt the discussion because they suspect that the crimson tide gives us secret powers and we can read their thoughts and shit. Anyway, whether or not it's time for the Red Wedding, curl up with a hot-water bottle and read *Flow: The Cultural History of Menstrua-*

tion by Elissa Stein and Susan Kim (not to be confused with *Flow: The Psychology of Optimal Experience* by Mihaly Csikszentmihalyi, which is also an awesome book, but not about periods). If you've ever wondered what kind of stuff had to be stuck up there before modern tampons, look no further. It's worth a browse for the kitschy old maxipad advertisements alone. You could also learn more about periods the way everyone else did: *Are You There God? It's Me, Margaret* by Judy Blume.

"I have a friend who needs consoling. Gonna grab some Cheetos and wine and head over there."

By all means, carry through with one facet of this falsehood and get some Cheetos and wine. You deserve it. And, being that you're the type of pal who lies about having to console a nonexistent buddy, you deserve some books about true friendship as well. *Quicksand* by Steve Toltz is a satire about two bros — one a writer and police officer, the other a con and the writer's muse. You're never going to be disappointed with *How to Make an American Quilt* by Whitney Otto. And by that I mean *I'm* never going to be disappointed with it. Everyone needs a story of intergenerational women making a quilt once in a while. Rent the movie too, with Maya Angelou, Anne Bancroft, Winona Ryder, and a young ga-ga-ga-gorgeous Johnathon Schaech. Then you can read the book, watch the movie, get drunk, and pass out in your Cheetos and have a dream about doing it in an orchard with Johnathon Schaech and/ or Winona Ryder. Classic Saturday night.

"I went out last weekend and now I have this rash that keeps spreading."

Gross. But we sometimes go to great lengths to get out of being social. Also, this excuse might prompt your friend to tell you about his or her own experiences with rashes and then you know whom not to share bathing suits with anymore. Anyway, I've got a great rash book: *Black Hole* by Charles Burns. A graphic novel where all the teenagers in 1970s Seattle start getting strange, seemingly sexually transmitted, mutations. Sometimes, it's a weird bump they can cover up. Sometimes, all their skin starts peeling off. It's amazing.

"Can't get a sitter."

"Ack, sorry. Kids are fighting again," aka, I'm in bed reading *Lord of the Flies* by William Golding. "I'll be at a fiftieth birthday party where none of the guests are supposed to know the hosts are getting a divorce," aka, after bathtime, I'll be devouring *The Adults* by Alison Espach. "The teenage son I abandoned is driving across the country to visit me," aka, I'm totally lying — I'm just reading *Pacific* by Tom Drury, and it's terrific. "Uh, I just found out my twin sister was actually a chimp and I've decided to take a vow of silence," aka, I know you know I'm lying, but I don't care. I'm relaxing with *We Are All Completely Beside Ourselves* by Karen Joy Fowler.

Book Hookups:
MÉNAGE À LIVRE
(YOU AND TWO BOOKS)

Sunny days and bike rides. Rainy days and Netflix binges. Being drunk and eating tacos. Being sober and eating tacos. Some things are just better together. Guess what, this rule lends itself to books as well. When you've hit on a certain feeling that a good read can give you — especially a loveydovey feeling — you want to prolong it as long as possible. So prepare ahead of time and bring two books to bed tonight. Think of the following list as a personal ad for people in love with reading: a pair of books for every romantic feeling. All the novels are relationshipish, ranging from platonic to bow-chicka-bow-wow.

Meet Cutes: *Two Across* by Jeff Bartsch + *Attachments* by Rainbow Rowell

If you like to pal around with quirky, clever novels that have well-crafted hooks,

238

introduce yourself to these guys. In *Two Across,* Stanley and Vera meet when they tie for first prize in the National Spelling Bee. They decide to embark on a marriage of convenience so that they can live off the wedding-gift money while pursuing crossword-puzzle writing. Their plan goes awry. The book even offers romantic crossword puzzle clues. It is the nerdiest fun. Then move from crosswords to "cc:" with Rowell's *Attachments.* In this novel composed mostly of e-mails, Beth and her work-wife Jennifer trade e-mails about Beth's lackluster boyfriend on company time, even though they know the IT department scans all of their correspondence. Enter Lincoln, the IT dude reading the messages. Can you fall in love with someone through letters, even when they're not addressed to you?

Quickies: "We Didn't" by Stuart Dybek + *Paper Lantern: Love Stories* by Stuart Dybek

Lovin' in a hurry. First, polish off Dybek's short story "We Didn't," which is the quickest way (save actual quickies) to get that dreamy, lovey, regret-y feeling. Plus, you'll be like, "Why didn't anyone ever fucking tell me about Stuart Dybek?!?!" Then pick up *Paper Lantern,* Dybek's book of nine sensual love stories that are going to "seal the deal" on your digging this writer.

Forbidden Lovers: *Tulip Fever* by Deborah Moggach + *At Swim, Two Boys* by Jamie O'Neill

Being in a secret relationship is thrilling, but so time-consuming with the alibis and the finding sexy trench coats to disguise yourself. Better to take these two to bed instead. *Tulip Fever* is about an affair between the young wife of a wealthy older man and the painter commissioned to paint their portrait. Set in Amsterdam in the 1630s, when a lust for tulips had taken over the city, you'll feel all atwitter as the tension builds and the plots twist. Then travel to Ireland in 1916 for *At Swim, Two Boys,* the love story of Jim and Doyler, two teen boys who find their friendship blossoming into a romance in the years just before the Easter Uprising. Read *Tulip* first for the rush of titillating emotions, but finish with *At Swim,* where the feels come in big waves and threaten to overcome both characters and the reader.

Yin and Yang: *Love in Lowercase* by Francesc Miralles + *Desperate Characters* by Paula Fox

It's fun to see how two books can work with the same setup. So why not experiment? Sometimes, you want pillow talk; sometimes, you want the down-and-dirty talk. The aforementioned books have absolutely nothing in common, except this premise: a stray cat wanders in and upends a life or two. In *Love in Lowercase,* a lost feline wanders into the apartment of a linguistics professor, Samuel, and becomes the stimulus for his breaking out of an introverted shell and leading him back to his childhood crush, Gabriela. In *Desperate Characters,* Sophie, who lives with her husband, Otto, in 1960s Brooklyn, is bitten by a stray cat she tries to feed. Her fear of rabies spirals into underlying anxiety about her marriage and society, and the couple's posh high-society world begins to unravel. Both are pretty delicious reads for remarkably different reasons. *Love* for its absurdity and fun pop-culture references. *Desperate* for Fox's precise prose and finely woven plot.

Conjugal Visits: *Fates and Furies* by Lauren Groff + *The Happy Marriage* by Tahar Ben Jelloun

To be read ALONE in a room after a fight with your partner. By the time you're finished, you won't be angry anymore, because you'll have been schooled in the lesson of "two sides to every story." Books about marriages serve as a good frame for character development, and these two novels of husbands and wives don't disappoint. In *Fates and Furies,* a husband and wife who began their relationship with a passionate romance find disorder lurking. Groff tells the story of the messy underbelly of this marriage from each spouse's perspective. *The Happy Marriage* begins with a secret novel written by an unhappy husband who blames his marriage for his ailing health and career. Until his wife finds the book and begins journaling her own perspective. As in real relationships, none of these narrators are perfectly reliable, but they are reliably engrossing.

Modern Love: *The Argonauts* by Maggie Nelson + *The Intimates* by Ralph Sassone

Maybe you're looking to spend some time contemplating all the ways we fall in love today, beyond the traditional standards of man meets woman and procreates. There are so many things that a loving relationship can be, and these books exemplify just two of them, but very well. In *The Intimates,* Robbie and Maize meet in high school and, despite being drawn to each other, realize their relationship will not be sexual, so they decide to bond by becoming each other's "human diaries." The book is smart, sexy, and honest. Which is precisely what *The Argonauts* is — in Nelson's sort-of memoir (it's actually "autotheory," a combination of autobiography and social criticism), she describes her romance with her fluidly gendered partner, Harry Dodge, and their journey to building a family while weaving their way around conventional ideas of sexuality and maternity.

Libidoless Lovers: *How Stella Got Her Groove Back* by Terry McMillan + *The Uncoupling* by Meg Wolitzer

For when you're in bed and you'd rather be reading. Begin with the anthem for, well, getting your groove back, with McMillan's modern classic about Stella, a fortysomething divorcée who is successful in every area of her life, except love. Happily, a vacay in Jamaica and a younger man are about to help her out. This endearing and funny read remains popular for a reason. *The Uncoupling* (not a Gwyneth Paltrow memoir) tells the story about the women in one small town who one by one lose their desire for sex right after the high school drama department starts rehearsing *Lysistrata* for the school play (in the play, the women withhold sex so that the men will end a war).

Just Friends: *The Country Girls* by Edna O'Brien + *Ghost World* by Daniel Clowes

Sometimes, friendships can be way more dramatic than romantic relationships, especially when teen girls are involved. In *The*

Country Girls, follow Kate and Baba from rural Ireland to Dublin while they come of age and are alternately terrible and tender toward one another. When I read *Country* for the first time, I thought, "This is like Enid and Becky from *Ghost World* if they had lived in Ireland in the '60s!" Both pairs of girlfriends are growing up and scared of growing apart. Both books are sincere depictions of adolescence and how a young girl's first love can often be her first close girlfriend.

Love in a Pop Song: *Nick and Norah's Infinite Playlist* by Rachel Cohn and David Leviathan + *Signal to Noise* by Silvia Moreno-Garcia

I want to feel like I'm inside a pop song. Wrapped up in a Smokey Robinson note. That would be awesome. These books can bring me close to that feeling. In *Nick and Norah,* two strangers have the same taste in music but know nothing else about each other. That doesn't stop them from embarking on finding a secret live show performed by a band they love. Superfun read. And it's short so you can finish it and jump right into *Signal,* Moreno-Garcia's novel about three friends in Mexico City in the 1980s who discover that their love of music can actually change lives. One of the friends, Meche, can cast spells using music. Fantastical, vivid, angsty, and full of heart.

Getting Kinky (with Books):
> *Sex Criminals, volume 1: One Weird Trick* by Matt Fraction and Chip Zdarksy + *Mr. Fox* by Helen Oyeyemi

Get crazy and play around with some literary tropes. The graphic novel *Sex Criminals* is a traditional superhero comic, right? Except that the superpower our heroes Suzie and John possess: they can stop time when they orgasm. Obviously, they decide to use this power to rob banks. So far, three volumes of this series have been published, and they are a lot of uncouth fun. *Mr. Fox,* on the other hand, a novel that cleverly plays with the fairy-tale form, depicts a writer, Mr. Fox, whose muse comes to life and puts him inside his own stories. Upset that her husband is out traipsing through fairy tales with his muse, Mrs. Fox throws herself into the stories as well, and the writer must choose between his dream woman or the woman of his dreams. With these two books, you will definitely be learning some new moves, at least as a reader.

Literary Lovers: *Extremely Loud and Incredibly Close* by Jonathan Safran Foer + *The History of Love* by Nicole Krauss

Okay, they broke up. But for a sweet moment, the two great minds Foer and Krauss were married and each writing books, and the similarities between the novels they released provide a keen insight into the symbiosis of two literary minds in love. I like to picture them sitting back-to-back, typing furiously, with little warm vibes steaming off of their beautiful brains and melding together. Both novels take place in New York and feature cerebral children and links to a sad past in Eastern Europe. Both are awfully charming and sentimental.

Jealous?: *Endless Love* by Scott Spencer + *Was She Pretty?* by Leanne Shapton

Obsessive desire feels awful in reality, but reading about it is totally thrilling. The above novels are a match made in insecure heaven. In *Endless Love,* which manages to be both sexy and eerie as shit, David and Jade are in the thrall of a passionate affair, but when Jade's family forces their separation, David decides to start a "safe" fire to their house and rescue them as a way to win back Jade's affection. I think we can all agree ahead of time that David is going to be an unreliable narrator. In Shapton's illustrated rumination, *Was She Pretty?*, we look into a list of people's ex-lovers through the eyes of their current partners.

Mature Audiences Only:
Assisted Loving: True Tales of Double Dating with My Dad by Bob Morris + *Julie and Romeo* by Jeanne Ray

Everybody talks about the fantasy of growing old together, but it's not often we can find books about finding love when you're already old. Turns out when you can find them, they're pretty entertaining. Begin with *Assisted Loving,* Morris's convivial memoir about helping his father, recently widowed and in his eighties, find a companion while looking for love himself. Then for a little dessert read *Julie and Romeo,* a romance about two senior florists whose families are longtime enemies (hence, kitschy title).

Opposites Attract: *Persuasion* by Jane Austen + *The Fling* by Rebekah Weatherspoon

I just want to see these two books sitting on top of each other on a nightstand. How can a romance novel change in nearly two hundred years? A LOT. But the thing is, both Austen's subtle wait-the-entire-novel-for-a-modest-declaration-of-love and Weatherspoon's raunchy girl-on-girl affair (while male fiancé is away) give you the tingles. *Persuasion* is Austen's last and my favorite of her books — a heroine who waits SEVEN YEARS after breaking her engagement with a man she truly loved to come back into his arms. In contrast, the heroine in *Fling* waits about one hot minute after her boyfriend gets on a plane to sleep with her female personal trainer. But a fling turns into an affair (and an affair turns into love). Surprisingly, both novels explore similar themes of society's opinion of and power over our romantic decisions as well as main characters trying to decide if they should take the risk and fall in love.

READIN' NERDY:
BOOKS ABOUT LIBRARIANS

The world needs a list of books that feature librarians and, dammit, I'm going to give it to you. Not because it's a scant list. In fact, it's pretty easy to find novels with librarian characters, especially in the romance and mystery genres. But those books tend to resemble a literary game of Fuck Marry Kill. Authors always seem to want to rescue female librarians from spinsterhood and turn us naughty or bludgeon us to death. Or both! See: *Murder of a Stacked Librarian.* THIS list includes novels about librarians that aren't typecast as Vamp or Corpse. The only thing the librarians in this list have in common is that they work at libraries. And wear cardigans. That stereotype is true. We are always cold as shit because we have snake blood and forked tongues. Our shushes are really lisps. Anyway, onward.

The Uncommon Reader by Alan Bennett

This novella features a librarian so against type, he's not even technically a librarian. Even so, Norman the kitchen boy carries out the two most important tasks of a librarian: hooking readers on books and defending the freedom to read. The reader he hooks is none other than the queen of England, who meets Norman when she stumbles across a mobile library near the palace. This short read packs in a great deal of humor and heart. The queen is portrayed in a lovable light, but as she finds herself, thanks to Norman, becoming an insatiable reader, those around her mistake her new abstraction for a growing feebleness of mind. Absurdity ensues.

In the Stacks: Short Stories About Libraries and Librarians by Michael Cart

Many writers credit the public library with their education and love of reading, so it's not surprising that the great ones tend to have a library/librarian story up their sleeve. But you will be pleasantly surprised at Michael Cart's compilation of short stories about book slingers and their shelves. *In the Stacks* includes the work of, as my father would say, the Upper-Crusty-Mucky-Mucks of fiction writers: Jorge Luis Borges, Isaac

Babel, Alice Munro, Italo Calvino, Ursula K. Le Guin, Ray Bradbury, Lorrie Moore, and John Cheever, to name a few.

Jade Dragon Mountain by Elsa Hart

Librarians are bad-ass investigators, and this novel features one such research expert, Li Du, the wandering exiled imperial librarian, who, in 1708, enters the last town in China before the Tibetan border to find it full of crowds. The people are gathered for the arrival of the emperor, who is visiting to "orchestrate" an eclipse of the sun. And when a Jesuit astronomer is murdered in the home of the local magistrate, Li Du is compelled to solve the crime. Lots of interesting historical details, including mysterious libraries and literary references.

In One Person by John Irving

Not only is John Irving an all-around boss writer, he is also a gallant writer who explores the dark corners we've ignored in ourselves and society. In *In One Person,* protagonist Billy is an aging bisexual man looking back on his life and romances, including his adolescence in the 1950s and '60s, when he harbored a deep (and eventually fulfilled) longing for Miss Frost, the older transgender local librarian. It's a gripping read (and I'm using that word to mean captivating, but also with a little wink because there are a lot of sexy parts in it).

Love and the Art of War by Dinah Lee Küng

Couldn't get a cuter premise. Jane is a librarian in London who signs up for a seminar about saving her marriage. Except she accidentally wanders into the room across the hall, a seminar for businesspeople about using Sun Tzu's *The Art of War* in their offices. The professor convinces her to stay and use the principles she learns to win her wayfaring husband back. I mean, who wouldn't want to read that? Fun and witty. When I read lighter fiction, I like it to take place in England, because it's much more enjoyable when you read it to yourself in a British accent.

When Tito Loved Clara by Jon Michaud

Michaud, former head librarian at *The New Yorker,* wrote this wistful novel about Dominican American Clara, who tries and often fails to escape her childhood and complicated family. Clara, a librarian, marries fellow librarian Thomas (that's two librarian protagonists if you're tallying), and, together with their son, the family lives a calm if not exactly happy life. Cue entrance of old boyfriend Tito, who seems not to have matured in the fifteen years since he and Clara parted. His love for her remains unchanged as well and could be the undoing of the both of them.

The Time Traveler's Wife by Audrey Niffenegger

I dig it when male librarians are represented in literature, although I'm jealous that they tend to get more heroic plotlines than the females. Henry is about as fearless and willful as any epic hero, though his only task is to stay alive through his involuntary time traveling long enough to get back to his equally brave and seriously stressed-out wife. Also, he has to try and not get fired at his library job since he frequently ends up naked in the stacks after he time travels.

The Book of Speculation by Erika Swyler

This surreal novel moves back and forth between the past and future of one family. In the present, sad Simon from Long Island works as a librarian and lives alone in his crumbling family home. One day he receives an old book filled with odd pictures of carnival life and discovers his grandmother is mentioned in it. The narrative that takes place in the past is the story of those mentioned in the mysterious book — wandering carnival acts who are somehow linked with Simon's maternal history and possibly a strange family curse. And there's merrrrrmaaaaids!

Mayumi and the Sea of Happiness by Jennifer Tseng

I mean, another sad librarian. We're not all so disenchanted, but, c'mon, who wants to read a novel about a happy library employee whose only character development is getting pumped for the interlibrary loans? Mayumi is a librarian on a coastal island off New England who awakens from her routine life as a wife, mother, and career gal by obsessing over a seventeen-year-old male patron. After months of fantasizing, she embarks on an affair and also begins a friendship with the young man's mother (don't ask). The obsession only intensifies, rendering the life that Mayumi returns to every day that much more oppressive.

Hard-Boiled Wonderland and the End of the World by Haruki Murakami

Oh man, this is a book by Haruki Murakami. There is a librarian in it. That's all I can say for sure and also should be enough information to make you want to dive into this complex, heady, genre-bending tide of "Whaaaa?????" But I'll give the plot summary a whirl: with each chapter in this novel, the reader alternates between two worlds. The first: reality (sort of), where our hero, a data processor, encounters an evil-genius-professor type who messes with his brain. He can steal sound. The second mythical world is weird as shit and that's where the library is and it has unicorn skulls in it. There. I tried. Read it, though.

FALLING DOWN
THE RABBIT HOLE:
BOOKS THAT LEAD TO MORE BOOKS

Sometimes, a book can take you on a journey far beyond the story itself. Sometimes, one thing leads to another and, six months later, you're reading about the existence of telepathic bionic soldiers and someone asks you how this obsession came to be and you're like, "My friend Kyle gave me *The Tipping Point* by Malcolm Gladwell." And they're like, "What the?" And you're like, "Seriously, shut up. I'm reading." It's the best kind of downward spiral. Let me take you on one now . . . (Cue: Gene Wilder as Willy Wonka: "If you want to vieewwww Paradise . . .")

The Virgin Suicides by Jeffrey Eugenides
At this point in the book, you may have gathered that I'm more than curious about the work of Mr. Eugenides and *The Virgin Suicides* is my ab fave, so let's start here. A dark story about the five gorgeous and

depressed Lisbon sisters, told from the brilliant point of view of the group of boys that grew up watching them, adoring them, and losing them. Every line is perfect. I. Just. Can't. Even.

We Agreed to Meet Just Here by Scott Blackwood

Needing to feed the beast for another lyrical beaut of a story about a community suffering from loss, Blackwood's slim novel is just the thing. This book is also told from the perspective of an entire neighborhood and features a beautiful young woman in danger. But there is another character, Odie, an old man who, in his youth, was the first person to find the bodies of the Peoples Temple at Jonestown and, later in the book, believes himself to be having conversations with Jim Jones himself. And then you're going to get a little taste of the details surrounding the Jonestown cult and mass murder/suicide and you're going to need to know more about that.

Stories from Jonestown by Leigh Fondakowski

The author of this book spent years compiling interviews with the survivors of Jonestown, most of whom lost family members in the massacre. Learn about their lives inside the Peoples Temple movement, which did not start out to be such a crazy movement, you guys. It was its own dark rabbit hole. You'll also hear about the lives of the survivors in the aftermath of the tragedy.

Going Clear: Scientology, Hollywood, and the Prison of Belief by Lawrence Wright

Now you're into the world of religious cults. You're welcome. I won't kid myself into thinking you're not going to be reading pretty much every other book ever written about Jonestown first, but when the Jim Jones well runs dry, you've got L. Ron Hubbard. And he'll fill the evil crazy hole in your heart. Wright, an investigator, gives it all up in this exposé, including that this religion was started by a man who believed he came to know celestial secrets under the gas at the dentist and some good stuff about Travolta.

Gold Fame Citrus by Claire Vaye Watkins

If you're starting to bum out a bit from the real-life corruption and power-hungry prophets but you're still aroused by the Hollywood glitz and the lure of the soothsayer, take your jaded heart over to the fiction section and pick up *Gold Fame Citrus*.

In the near future, serious drought has left California deserted, save mostly for criminals and hippies. A young couple is camping out in a starlet's abandoned mansion in Laurel Canyon until events move them to leave L.A., with a small child (okay, a stolen child and this is another connection with *Going Clear* because Hubbard is said to have abducted a baby as well). They stumble upon a colony of people living in the massive and otherwise abandoned sand dunes caused by the drought, led by a prophet figure. That's all I'm going to say. So good. This is one of those books you're going to ignore your family and friends for, so go ahead and set out some snacks ahead of time. Gah, it's good. I pictured Eva Mendes and Ryan Gosling for the young lovers and Brad Pitt as the mysterious sand nomad, if anyone's asking.

No One Is Here Except All of Us by Ramona Ausubel

"Oh, hello, I just finished *Gold Fame Citrus* and I'm looking for another excellent, lushly written fantastical novel that features a community of people who have cut themselves off from society in response to a major world event."

Can. Do. Ausubel's novel takes us backward in time instead of forward, to 1939 in Romania, where a small Jewish village knows the war is closing in on them. So they decide to willfully ignore it. All of it. All of history. They all give each other new jobs and spouses and children. And although it certainly isn't an uplifting book, it's a very loving book and it will make you Think and Feel.

Everything Is Illuminated by Jonathan Safran Foer

Another novel with the qualities of a fable and a connection to World War II, *Everything Is Illuminated* features a young man who travels to a village in Ukraine seeking a woman believed to have saved his grandfather from the Nazis. This story thread is paralleled with the narrator's reimagining of the past history of the village. Foer touches on themes familiar to *No One Is Here Except All of Us,* but his story has some definite comedy to it, in case you were feeling like you're going to need therapy if things don't start looking up on this book journey.

***Running Away to Home: Our Family's Journey to Croatia in Search of Who We Are, Where We Came From, and What Really Matters* by Jennifer Wilson**

Switching back over to nonfiction but keeping the Eastern European vibe and the interest in family history, join Wilson as she writes about how her family picked up and left their indulgent life in Iowa and moved to the village in Croatia where Wilson's ancestors hail from. The author's search into her Croatian ties and the experience of the family's sincere effort to live a more simple life make for an endearing read. They eat a lot of sausage. Whether that grosses you out or makes you want to nosh on some kielbasa, be prepared.

***Great Plains* by Ian Frazier**

Once you've whet your appetite for travel writing with a soft spot for history, you're going to get wanderlust, and Frazier's *Great Plains* is a great book to wander into. Frazier travels by rusted-out van across the Great Plains, visiting historical sites like the location of the murder that sparked the true-crime genre, *In Cold Blood,* and regaling readers with the history of the prairie people, from Crazy Horse to Bonnie and

Clyde. You'll be like, "Dang, road trip anyone?"

The Disenchantments by Nina LaCour

But if you're broke or don't have any friends to go road trippin' with, read the charming YA novel *The Disenchantments*. I love "we just graduated" stories, and this one doesn't disappoint. Colby tags along with his BFF and unrequited love, Bev, and her all-girl band on a summer tour. Bev's growing distant, they're really into the rock group Heart, they're angsty and selfish and confused and sad all at once. It is an excellent summer-afternoon read. But either unfriend your high school crush ahead of time or don't drink for a few days after you finish in order to avoid leaving an embarrassing Facebook message for them: "I was so in love with you! And you Didn't. Even. Care."

And if, after all that, you're hankering for another novel about teen boys longing for distracted and disturbed teen girls, it's probably about time you read *The Virgin Suicides* again. Except this time when someone asks you if you like it, you can say, "Yes, and now I know all about the Peoples Temple and kielbasa and John Travolta and the Nazis."

HE'S JUST NOT
THAT INTO LITERACY:
TURNING YOUR LOVER INTO A READER

> If you go home with somebody and they don't have books, don't fuck 'em.
> — John Waters

Man, don't do it with someone that doesn't enjoy reading! What are you going to talk about after? But all is not lost. Librarians turn people on to reading every day, and we are fervent believers that there is a book for everyone. Let's walk through the process of turning your lover from easy to easy reader.

Casually ask your partner what the last thing they read was and refer to the following typical responses for reading advice. We'll get through this together.

"Hunger Games. *I read the whole series in three days.*" (*Or* Fifty Shades *or* Twilight, *etc. etc.*)

Okay, this is promising because this answer means your lover already loves reading. They're just too lazy to seek out books for themselves. Help them realize that they don't have to wait until the next movie franchise to enjoy a thrilling read. This type of book snacker likes drama and fast pacing, and they probably prefer the momentum and deluge effect of a series. Some obvious recommendations are the Divergent series, by Veronica Roth, and *The Lunar Chronicles* by Marissa Meyer. But if you think they're ready to branch out of the young-adult genre, try *Annihilation* by Jeff VanderMeer (first of a trilogy), or any of Gillian Flynn's addictive novels. Both VanderMeer and Flynn have books that have been turned into movies but are so good that your partner might forget about films altogether and dive into the written word.

"An Onion article about dolphins."

This says to me humor in short savvy doses that give off a teeny jolt of superiority. There are definitely books for this sort of reader. *The Onion* publishes a book every once in a while. But to further expand your honey's horizons, hand over the smart 'n' snarky essays in *Fraud* by David Rakoff. For more hilarity and even some blushing, give them Jenny Lawson's dark gut-buster of a memoir *Let's Pretend This Never Happened.* Also, either of the Sedaris siblings (David or

Amy) are a no-fail. I find something newly hysterical every time I read Amy Sedaris's *I Like You: Hospitality Under the Influence,* including the autograph I have on the title page, in which she wrote, "Annie — Drinking kills feelings."

"My mom's recipe card for Pungent Chicken Livers. Take a whiff!"*

First, there is a not-so-small population of people who sit down and read cookbooks like novels. It's a thing.

I recommend *Vegan Soul Kitchen* by Bryant Terry and *Momofuku Milk Bar* by Christina Tosi for that. An appetizer of books. Next, a steady diet of foodie writing. Start with Lucy Knisley's *Relish,* the memoir of a food lover in comic book form (with recipes!). If they're open to that, gently guide them into fiction with *Like Water for Chocolate* by Laura Esquivel and its modern counterpart, *The Particular Sadness of Lemon Cake* by Aimee Bender, both magical novels about characters who are able to transmit emotion through food. With any luck, the message

* This person's mom got the Pungent Chicken Liver recipe from page 233 of the 1969 edition of Evan Frances's *Family Diet Book,* so don't let her tell you she "threw it together."

their next savory dish will express is "What do you know, I *do* like to read!"

"The liner notes to [insert latest vinyl album that still has liner notes]."

If you're going steady with someone who likes the dirt on his/her favorite musicians, they will love reading *The Dirt* by Mötley Crüe and Neil Strauss. Even if they've never listened to "Dr. Feelgood." Doesn't matter. It's vulgar and salacious and a great opener into the musician-biography section. Once interest has been piqued, let flow the river of pop-music profiles. You can't go wrong with anything by Peter Guralnick. Also *Le Freak: An Upside Down Story of Family Disco and Destiny* by Nile Rodgers is fascinating. Rodgers is the man behind most of the music you love, including David Bowie, Madonna, and Daft Punk. Also *Hunger Makes Me a Modern Girl,* the memoir of Sleater-Kinney's guitarist, Carrie Brownstein. After reading these, your partner's music-book pile is going to start filling up on its own.

"Vogue. *Like the magazine.*"

For those moved by style but "meh" on substance, the perfect book to balance the two out is *Women in Clothes* by Sheila Heti,

Heidi Julavits, and Leanne Shapton. Okay, it's over five hundred pages BUT it's so dapper and divided into short, magazine-like features — interviews, photographs, and conversations about the meaning behind what women wear. You could also try artful books that your sweetheart will gravitate to for their briefness and chicness. Try author/ designers Maira Kalman or Robert Klanten. If nothing else, your main squeeze can pose with them in front of an urbanish wall for their fashion blog.

"The Unbearable Lightness of Being." (*You know for a FACT this was twelve years ago for a college course, but Lover Boy/Girl brings it up at parties like they just read it.*)

The only social perk readers get (the antisocial perks are aplenty) is seeming well-read at parties. If your escort is trying to earn cred with the same impressively dense novel at soiree after happy hour after potluck for a full decade, that is a mockery of your carefully assembled Have Read/ To Read piles, and you may need to reevaluate your relationship. If they start talking about books *you* read at parties, using details you told them while seemingly ignoring you? Instant dismissal. If you decide to stick around, demand to hear their thoughts on *Infinite Jest* by David Foster Wallace. At 1,079 pages, it will give you something new to talk about and also serve as penance for spending twelve years without reading.

"*Um, the last episode of* Scandal?"

True, *Scandal* is a television show and not a book. But you would be surprised how often librarians have to give book suggestions based on a movie or TV show that people like because it's been so long since they read a book that they no longer have a reference for what they would like to read. That's okay. We don't judge. We pity. And then we load their asses up with great books. So if your betrothed spends more time in front of the boob tube than the word cube (I tried), give them a book that meets the same criteria. For *Scandal* lovers and most reluctant readers, a great gateway genre is true crime. These books are usually quick, fascinating, and leave you wanting more. Start with *The Man in The Rockefeller Suit* by Mark Seal, about the dude who impersonated a Rockefeller and learned how far bravado and mediocre lies will get you (pretty far, turns out). *Tinseltown: Murder, Morphine and Madness at the Dawn of Hollywood* by William J. Mann also doles out a hearty dose of scandal. After they're hooked, they can transition to the classics of the true-crime group (*In Cold Blood, Helter Skelter*) and subsequently into fictional crime classics.

"My fantasy-baseball draft picks."

Jocks don't get a lot of credit for being readers, but judging by the number of sporty books out there, I'd say that's a misconception. Especially baseball books. There are A LOT of baseball books, but, like Dolly Parton says in the movie *Straight Talk,* the world could always use more baseball books — if they're good. And Dolly is always right. So grab an autobiography of your friend's favorite player or an historical account of the year their team won/almost won the World Series. When your friend has finished those, slip in *Shoeless Joe* by W. P. Kinsella (the novel *Field of Dreams* was based on), and *The Art of Fielding* by Chad Harbach, a baseball book but also a multilayered novel about relationships. Your dearest will soon be addicted to reading. And instead of being annoyed that they don't like to read before bed, you're going to be pissed because they took your Joyce Carol Oates.

"I've just recently shamed myself by binge-reading mommy blogs."

Hug this person. They are probably going to resist taking the time out to read because they suddenly feel the need to get rid of all the sugar in the house and make their own dry shampoo and buy a sensory development table and cry themselves to sleep. But they need to take a fucking chill pill and read a book that isn't about parenting. *I Think I Love You* by Alison Pearson is a delightful romp of a book (in a British accent!) about a woman who finds her teenage dream of meeting David Cassidy coming true when she least expected it. I also recommend *Pride and Prejudice and Zombies* by Jane Austen and Seth Grahame-Smith for this type of reader. Just enough absurdity to make you forget your worries for a few hundred pages.

BLANK STARE

Your boo can't even recall a title of a book he or she has *heard* of and so can't lie about having read anything. This is a desperate situation and calls for professional intervention. Drag them into the library, get them a damn card, and hand them over to the reference desk. A librarian will know how to ask the right questions to illicit an actual response and send your lover home with a stack of books (probably audiobooks — baby steps) to try out. And if they actually cannot read (which is unfortunately more common than you would think), the library can help with that as well.

Literal Reality:

GET OUTTA THIS WORLD, GET INTO THESE BOOKS

Do you ever feel so wrapped up in a story, it's like you're more in that world than your own? Gah, I love it when a book's atmosphere is so intense that you have to look up to remind yourself where you are. You're there in the book, rummaging through someone's empty house looking for food after a near apocalypse and then suddenly your husband's like, "Pizza's here!" and you know you're at your house, at your own dinner table, telling your kid to sit still and asking your husband how his day was, but part of you is still in the book and part of you is still an apocalypse survivor. That's why you're hoovering the pizza like you haven't eaten in days. JK, that's just how I eat pizza.

Anyway, although sometimes other worlds in the following books may still be on earth, they all have the ability to suck you into their unique environment. Just remember to listen for the delivery guy.

Tales of Moonlight and Rain by Ueda Akinari (translated by Anthony H. Chambers)

This book of Japanese occult stories was written in 1776 and pulls you into dream worlds full of spirits; half-human, half-animal beings; and the occasional demon — all the things that come out to play on rainy, moonlit nights. Find a dreary evening and some blankets to hide under and give *Ugetsu Monogatari* a try (that's the transliterated Japanese title, if you want to sound smarter).

Parable of the Sower by Octavia E. Butler

In a dystopian future, in a gated community near what had been L.A., fifteen-year-old Lauren is afflicted with the superpower of "hyperempathy" — an ability to feel the emotional pleasure and pain of others. This means she can't ignore what the world has become or the suffering of those around her. Lauren creates a sort of "religion of change" called Earthseed, and, after her community and family are destroyed, she begins to move on with her belief in the future and her new companions. This is a two-book series and will rope you into reading more by Butler, the grand dame of science fiction.

Childhood's End by Arthur C. Clarke

Dystopian fiction sometimes feels like too much or hits too close to home. Good thing there is such a thing as *u*topian fiction and *you* would do yourself some good by beginning your idealist adventure with Clarke's novel, often heralded as the closest thing to a perfect science fiction book that exists. As the story goes, you know, aliens descend and take over every major city on earth. But they end up being pretty cool about the whole thing and eradicating a lot of human foibles. The book asks, even though the aliens have cured earth of disease and poverty and war, is this new way of life all it could be?

Little, Big by John Crowley

Fairies! In this hazy, magic novel, the Drinkwater family lives in a house called Edgewood for four generations — a house on the border of a fairy world. Don't be turned off by the fantasy stuff. The novel is more of a family saga with some flare and really lyrical language. It's been compared with, and actually quotes, *Alice's Adventures in Wonderland* (aka drug references, a chick named Alice).

House of Leaves by Mark Z. Danielewski

This novel is classified as ergodic literature. I looked it up. I don't get it. It doesn't matter. Still amazeballs, but aggressive with the footnotes! This is a horror/love tale told from the perspective of a tattoo guy and the perplexing manuscript he finds in a dead man's apartment. You will not only escape your life with this one, you're going to have to call in sick for a few days. Don't let my husband sit next to you when you're reading this because you need to focus. He will, for sure, interrupt several times and end up turning on *Fool's Gold* with Matthew McConaughey and Kate Hudson. It's really hard to read ergodic lit with sexually charged banter and a Bob Marley soundtrack playing in the background. This much I know.

Hotel World by Ali Smith

The haunting language and stream-of-consciousness narrative in Smith's novel is what invites the reader to believe they are part of this world, sometimes as one of the living and sometimes . . . not. (Has it been long enough that "I see dead people" is funny again?) Five female narrators who all have one fancy English hotel in common let us into their thoughts about life and death's big questions. A dark 'n' quirky read.

Dahlgren by Samuel Delany

Seventies sci-fi American magical realism metafiction. Does that make you scared or excited? Weird shit is happening in one Midwestern city. Like, time is passing differently for different people, there's a wanderer who doesn't remember who he is, there's no money or electricity, but there's a lot of sex. This book is hard to get through and is not for everyone. "Not for everyone" actually seems to be the unintended subtitle of *Dahlgren* if you read the reviews. But it's also the ultimate "live within a book" kind of book. So give it fifty pages. Then give it all 816. What, have you got something better to do? You're reading *this,* aren't you?

What I Didn't See by Karen Joy Fowler

Too fickle to commit to one dreamworld for the afternoon? Read Fowler's short-story collection, which runs the gamut of subject matter and protagonists — from a bratty teen to a woman suffering from Alzheimer's to John Wilkes Booth — but sustains a fantastical (mostly dark) thread throughout. Fall into strangely beautiful scenes, one after the other, and return to reality by bedtime.

Never Let Me Go by Kazuo Ishiguro

A "first love" story if you met your first love at an artsy boarding school called Hailsham that you weren't allowed to leave. Kathy, our heroine, is assigned the job of a "carer" (you'll figure it out). As an adult, she reconnects with two of her fellow classmates from the past, and the three probe further into just how they came to be at Hailsham in the first place and how it alters the course of their lives. Elegant and atmospheric — let go and let it move you. This is a story that has to settle inside you for a bit before you pick up another book.

Station Eleven by Emily St. John Mandel

Mandel's novel is powerful and builds a detailed world while remaining a gripping read. In a not-so-distant future, 99 percent of the population and the world's technological progress has been wiped out by a flu epidemic. Wait, though. It's not about *that.* It's about how people move on, and how they find comfort, companionship, and hope for their future. And there's a crazy prophet in there. I can't put down a book once I know there's a crazy prophet involved. The postflu world Mandel creates is beautiful and oddly peaceful at times, as much as a world filled with the artifacts of a former way of life can be. Unfortunately, I don't wash my hands nearly enough to fantasize that I would have been a survivor.

BLIND DATE:

GOOD BOOKS
WITH BAD COVERS

There are two things everybody says not to do, but everybody still does anyway: peeing in the shower and judging a book by its cover. Am I right? Guys?

Well, anyway, think about how difficult it is to come up with ONE image that totally evokes an entire book's identity. Nearly impossible. I can't even think of an example of a book jacket that 100 percent captures its insides. Except maybe this book. If I get my way, the cover of *this* book is going to be a knockoff of the *Sgt. Pepper's Lonely Hearts Club Band* album, except I will be all four of the Beatles' faces and all of the decoupage heads are going to be famous authors (plus Jared Leto) crying because I've weeded them from the library. Also I promised my cat, Barb, she'd be on the cover. That, I think, will give people a fairly good idea of what I'm going for here.

The following book covers aren't actually

bad (psst — some of them are), they just don't do their pages justice. So if you chance upon some of these books and you like their summary but they don't look so hot up front, give 'em a gander anyway. They are more than assaults the eye.

Jane Eyre by Charlotte Brontë

Not every edition, because there have been a gabazillion, but, man, there are some pretty heinous *Jane Eyre* book jackets floating around. Most covers tend to fall flat in conveying the governess's persevering spirit. If someone were to call Book Cover Jane up on the phone and ask what she was doing, she might coolly reply, "Just chillin', you know, doing some needlepoint, wandering around the gardens, braiding my hair, whatever, not at all worried about that strange old lady in my attic." What's worse than these covers, though, are the ones that make this nineteenth-century literary classic look like the next title in the *Goosebumps* series, or the cover with Jane decked out with red nails and lips, holding a rose. *Jane Eyre* is many things — spooky, thrilling, passionate, and tender — but a rom-com it is not. If you haven't read this book, or if you haven't read it since you were required to read it (and didn't really read it), give it another try and embrace whatever cover girl you end up with, be she haggard, sultry, or dead in the eyes. They're all the same on the inside.

The Martian by Andy Weir

It got a cool cover. Then it got a movie cover (blech, movie tie-in covers). But before all that, this funny, adventurous novel about an astronaut-botanist stuck on Mars had kind of a boring cover. It was just — Mars. Which, for previously self-published author and NASA enthusiast Weir, was probably really exciting. But I didn't know how exciting Mars could be until *after* I read Weir's book. When I first looked at the cover, it said to me, "Probably about an alien war." I read it anyway because my big sister told me to and I have to do what she says. If you passed *The Martian* by because it looked too fantasy, or too sciencey, or too Matt Damony, you missed out on a book that is truly entertaining and informative, and also you skipped book club out of guilt, and they probably talked about you behind your back.

Two Guys from Verona: A Novel of Suburbia by James Kaplan

This is a novel with a lonesome mood, about two high school pals who are living very different lives twenty-five years after graduation, yet each feels sorry for the other. It gives you a feeling throughout that a dark force is looming, in a really good way. So, I kind of get the hovering clouds on the blue cover, but it really makes this terrific novel about friendship and regret look pretty "meh." And the font bothered me. Why italicize the *from*? It's ridiculous that four italicized letters might keep someone from enjoying this novel, but that was almost my fate. What prompted me to open the book jacket was the subtitle, "a novel of suburbia." Characters from the suburbs have a specific malaise that I'm drawn to as a reader. Anyway, once I opened *Two Guys,* I consumed it entirely. But I almost didn't. So now I (mostly) don't let a book's font rule my reading.

Dandelion Wine by Ray Bradbury

I'll give the cover artists this: *Dandelion Wine* is hard to define. It can't be pigeonholed, which is what makes it so sweeping and tremendous. *DW* is not Bradbury's typical science fiction tome. It's not really a true coming-of-age story either. It's not a quirky townspeople tale. There are chapters that make you want to roll around in green fields and revel in your aliveness, and chapters that compel you to lock your doors and hide in your bathtub with your kitchen knives. It was probably a real bitch to come up with a book jacket for this masterpiece. The 1957 first edition — a field of dandelions — is vague enough that anyone with a passing interest and a slim recognition of Bradbury's name would probably be interested enough to pick it up. After that, the covers vacillate between creepy beautiful and weird as shit. By weird as shit I mean a kid in a serious jean jacket oblivious to a dandelion/jellyfish thing sucking his head, or a wine bottle with skulls and jaws and scissors and butterflies coming out of it.

The Wilds by Julia Elliott

Elliott's book of short stories features a beautiful cover of a woman with a cat's tail and a tree growing out of her head. It's intriguing and very well done. Still, I don't think it represents how nuanced and diverse the stories compiled inside are. It looks like a book of strange fairy tales. And there is most assuredly some strange shit inside. There are medical spas with flesh-eating procedures and pirates. There are old ladies with robot legs. But there are also girls and women very much rooted in the realities of their existence. The stories are expansive in subject matter and deal with complex relationships and emotion. Please don't mistake *The Wilds* for a book of fantastical fables and put it back on the shelf. We don't like when patrons reshelve the books on their own.

Mink River by Brian Doyle

I was so put off by this cover that I started thinking there had to be something I wasn't understanding. A crow Photoshopped on top of some water? But the title is *Mink River* — where's the mink? I was confused. I'm still confused. In my confusion, I perused the first page and fell pretty hard for Doyle's novel about the inhabitants of a small town in Oregon and a talking crow (it's more than that, but I wanted to explain the crow up front). The language is lovely and begs to be read aloud, the characters are engaging, and there's a little bit of magic. I really adore it. Still, I'm glad I don't have to look at *Mink*'s face every time I sit down in my living room, because it feels like when someone you're dating gets a terrible haircut and you kind of don't want to make out with them anymore.

The Last Picture Show by Larry McMurtry
Specifically, the 1967 paperback published by Dell. The novel, which you may have ignored because you think of McMurtry as a "westerns" guy, is an epic coming-of-age about teenagers in rural Texas in the 1950s. Even the older characters in the town are stunted by inertia. Everyone is yearning for something. It's a spare and melancholy story. But you will not infer this from the 1967 cover, which reads, in big block letters: "The Savage Lusts of Adolescent Innocence" and pictures a bikini-clad lady leaning up against one of the guys from *The Outsiders,* in front of a sunset. It looks like a pulp romance. A wild romp in a Texas town! And while there's a lot of sexual bravado in *Picture Show* (including some stuff with a farm animal you'll probably want to skim over), the vibe of the novel — regret and longing — is sort of a boner killer. Take this cover with a grain of truth. Come for the thrills but prepare for the heaving sobs.

The Cat in the Coffin by Mariko Koike
The cover of this book, a cat staring at a woman's foot on a straw rug, is just fine for

a book about — I don't know, a funny stoner story where the cat starts talking and it remembers where the main character put her wallet, but won't say, and they spend the whole book trying to find it so they can go to the health food store they both like. Koike's novel is MUCH better than the book I just described. But you might not know it. The story is really about Masayo, an au pair, and her charge, young Momoko, whose only friend is her cat. And Momoko's father, with whom Masayo has fallen in love. And the woman that the father loves instead. And an ensuing obsession. And some dark deeds. It's suspenseful. The hijinks are fatal. No one, to my memory, loses their wallet.

Recovery Reads:

A BOOK LOVER'S HAIR OF THE DOG

I don't like the phrase "guilty-pleasure read" because I don't think you should be ashamed of anything you read as long as it makes you happy and it's not, like, an instructional guide on how to steal and eat kittens. I also don't like that the genre of the "feel-good read" gives writers a pass to write quickly and shittily (like how I just made up a crappy word to describe lazy writing?) as long as they hit all the "feel good" points. Thus, I offer to you: Recovery Reads. These are the books to turn to when you're on the mend from a book that gave you nightmares or left you in a dark headspace and you need some lighter fare (but don't want to give up quality). Recovery Reads are also great books to read before bedtime, especially after you've watched the news and you feel like you're going to wake up in *The Hunger Games* arena.

Squirrel Seeks Chipmunk: A Modest Bestiary by David Sedaris

When I need a time-out from the heavy issues of our society, real or novelized, I like to pick up the most absurd thing within reach. Sedaris's animal fables of forbidden love, bureaucracy, and fad diets will make you laugh out loud and also put your worries in perspective. As with anything written by Sedaris, it's double the fun to hear him read his work and *Squirrel*'s audio edition also features Elaine Stritch, so, c'mon, you're not going to get better than that.

The Secret Garden by Frances Hodgson Burnett

Listen, don't be above going in the children's section when you need something to shake off your worldly adult woes. You can find a simple but still beautiful piece of work with an all-but-promised happy resolution and a chance to revisit some of your favorite characters (before you discovered Hermann Hesse and hardened your literary heart). *Secret Garden* is about another hard and sickly heart, two of them, actually, that are rejuvenated by a hidden garden in an otherwise lifeless English manor. It's a little word balm for your soul.

Dolly: My Life and Other Unfinished Business by Dolly Parton

This autobiography is also a soothing, easy read with substance. Dolly is undeniably smart 'n' sweet. Her childhood in the Great Smoky Mountains is the reading equivalent of sticking your bare feet in thick, cool mud (if you like the idea of that — and you will after you read her book). Also, fun fact, Ms. Parton met her husband, Carl, to whom she's been married for fifty years, outside a Laundromat on one of her first days in Nashville. No matter what twisted *Gone Girl* read-alike you're coming off of, Dolly'll melt down your skeptic's shield with her warmth and humor.

Whatever Makes You Happy by William Sutcliffe

The British tend to do wit well, and Sutcliffe is particularly good at bright and charming novels that are sure to cure your book blues. In *Whatever,* three adult men neglect to send their moms cards on Mother's Day, and the mamas get to talking. They decide their sons need to grow the F up and that the most efficient route to that goal is, obviously, to move in with them, unannounced. Can you imagine? Sutcliffe does, and even though he pokes fun at everyone, there's some serious humanity to this book. It's good to read after a *Game of Thrones* binge, when you're still in the mood for quirky family dynamics but without the incest and beheadings.

Advanced Style by Ari Seth Cohen

I mean, just look at these ladies — bold colors, huge glasses, and admirable aplomb. After pondering the existence of such evil characters as Annie Wilkes, Patrick Bateman, or Nils Bjurman, you'll feel better knowing there are people like this walking around in the real world. And there's a sequel: *Advanced Style: Older and Wiser.*

32 Candles by Ernessa T. Carter

The young-adult collection is also a great place to go for novels about real-life shit that doesn't necessarily spend four pages describing the button on a cardigan that's about to fall off because the main character's life is "hanging by a thread." In *32,* our heroine, Davidia, grows up in poverty with parents ill-equipped to care for her and a high school crowd that doesn't accept her. So she jumps on a truck and gets the hell outta there. Davidia, who finds solace in John Hughes movies, is looking for the kind of ending Molly Ringwald would get, and it's so much fun to follow her as she tries to reinvent and, eventually, accept herself.

Amy Falls Down by Jincy Willett

When you get into a cynical reading funk and everything you pick up seems to have the word "dystopian" in the summary, what you need is a book about someone who's just as much of a cranky misanthrope as you — and learns to love life again. The trouble is, it's hard to find a book with that premise that's not been touched by the pen of ole Pappy Sappy. (Pappy Sappy is an old man-fairy that sneaks into writers' offices in the night to do a little maudlin'n'. He lives at Nicholas Sparks's house.) You'll get none of that with Jincy Willett's darkly upbeat novel (that's a thing), *Amy Falls Down,* about an older writing professor who lives as a hermit and has stopped creating stories. Just before she's scheduled for a where-are-they-now type interview, she falls and knocks herself senseless on the birdbath in her garden and doesn't get her wits back until she's watching the interviewer drive away, having no memory of what was said. It's the most heartwarming concussion story not about football you'll read. But it's also hilarious to watch Amy's life spiral out of her tightfisted control after the interview

293

is published and she's hailed as a genius.

Nobody's Fool by Richard Russo

"He never considered life to be as tight as some people made it out to be, but it wasn't that loose either." It's a terrific line, describing the thoughts of this book's main character, the lovable grump Sully; but it also makes for a nice summary of this sometimes funny, sometimes aching, always spot-on novel. Sully has made mistakes in his life, and some of them, like sleeping with a married woman, he's still making. His bad knee keeps him from getting work, and the people he surrounds himself with generally annoy him. Moreover, his grown son seems to be taking after his old man more than Sully would like. Has his luck run out? Did he ever have any? Read this when you're down about mankind, and you may start to notice a roguish glint in the eye of the curmudgeon who bitches at you about your grass being too long.

The Sweet Potato Queens' Book of Love by Jill Conner Browne

This is my high compliment for Browne's series of hilarious books, where everyone's named Tammy and no one is holding their tongue: a man asked me once for a book recommendation for his wife, who was dying, and just wanted to laugh. And *Sweet Potato Queens'* was right where I led him. The women Browne describes, a clan from Mississippi who love to dress gaudy and march in parades as well as hand out advice and recipes freely, laugh a lot at themselves. But more than that, they *love* themselves. And they want you to love yourself and love the life you've got left. And eat more Pig Candy. And wear smutty clothes. And laugh. Even better on audio so you can hear Browne's southern drawl.

All Creatures Great and Small by James Herriot

I know this book is super old. It was written in 1972, and the original cover is a dude rolling up his sleeves while a bunch of random animals flee him. And he's looking down at the horse, whom he is inexplicably about eighty times bigger than, like "I'mma get CHA!" Anyway, that doesn't matter because no library will ever discard it because it is ALWAYS popular. People love this book. I know because everyone who checks it out stops at the reference desk to tell me how pleased they are that we have it, and they all have rosy cheeks and they cradle their copy like it's a broken locket and they're Little Orphan Annie — so it's got to be warming some cockles. Maybe people read the warm memories of the veterinarian from Yorkshire out loud to their pets at home? I say this because sometimes the books come back with animal hair in them. But give it a go.

Through Thick and Thin:
BOOKS FOR THE LAZY, THE LIVELY, THE LONG-WINDED, AND THE LETHARGIC

START HERE *if you tend to enjoy verbose texts or if you think someone rolling their eyes at you and sighing is a normal part of conversation:*

I have found some of my favorite books while wandering through the stacks, not really sure what I was looking for (aka: supposed to be weeding the fiction). Sometimes I look for book spines that appeal to me or authors with some version of my name. Often, when I'm feeling in the mood to gobble something up instead of pore over it, I simply scan for thin books I can absorb in an afternoon (aka: while I'm supposed to be budgeting). In the fall, when I'm feeling back-to-school-y, I browse for big, fat family dramas and epic generational works. As readers, we're not always in the mood for something specific, so a title that's "pretty long" or "pretty short" might be all we're looking for. I encourage you to wander in your public library this way. But, in the meantime, because I'm that good, I've compiled a list of books under two hundred pages and over five hundred pages for you to peruse. To enhance the feeling of a real shelf-browsing experience (and because I'm that lazy), I'll only give you a few keywords about each book.

So here it is, the short and the long of it!

BEGIN HERE if you're fickle (you'll know because your mood ring will be drab brown) and weren't even sure you wanted to read this passage:

Here are some long books and short books.

TINY TOMES, UNDER TWO HUNDRED PAGES

I'll just get this out of the way quickly, since I know you're in a hurry. Anything by Grace Paley is going to be short and dense as shit with the most resplendent, impeccable, yet unassuming language and images and stories that will burrow into your heart. Don't browse for her; seek her out.

Still here? Want to try poetry?

Poetry Primer Kit: *Good Poems* edited
by Garrison Keillor
Classic, Cream of the Crop, Being Hu-

man: *Strong Is Your Hold* by Galway Kinnell

New, Strange, Honest: *Dear Jenny, We Are All Find* by Jenny Zhang

How about short-story collections?

Short Story Primer Kit: *Selected Shorts* by PRI Public Radio International (This is an audio collection of famousish people reading famousish short stories live in front of an audience, available through your public library or through PRI's podcast on npr.org.)

Comedy, Neuroses, Slice of Life: *Do the Windows Open?* by Julie Hecht

Absurdist, Millennials, Absurdist Millennials: *One More Thing: Stories and Other Stories* by B. J. Novak

Satirical, Sharp, Off the Wall: *In Persuasion Nation* by George Saunders

Classic, Lyrical, Southern Gothic: *The Collected Stories* by Eudora Welty

Novellas, anyone? (Novella translates to "widdle baby novel.")

Biography, NYC 1980s, Artists: *Widow Basquiat: A Love Story* by Jennifer Clement

Fable, What If February Lasted FOR-EVER: *Light Boxes* by Shane Jones

Subtle Prose, Everyman, Food Service: *Last Night at the Lobster* by Stewart O'Nan

Coming of Age, Reason There Are Novellas: *Goodbye, Columbus* by Philip Roth

Minimal, 1990s Japan, Quirky: *Kitchen* by Banana Yoshimoto

WORTH THE WEIGHT, OVER FIVE HUNDRED PAGES

Poets, Latin America, Visceral Realists: *The Savage Detectives* by Roberto Bolaño, 592 pages

Nineteenth-Century New Zealand, Mysteries, Page-Turners: *The Luminaries* by Eleanor Catton, 848 pages

Classic, Sweet Revenge: *Count of Monte Cristo* by Alexandre Dumas, 1,240 pages (Yeah.)

Communists, Writer's Block, Feminist Crises: *The Golden Notebook* by Doris Lessing, 567 pages

Genetics, Love, Challenging Reads: *The Gold Bug Variations* by Richard Powers, 639 pages

Children, Memory, Murder, Throw-Your-Book-Across-the-Room-Suspenseful:

The Little Friend by Donna Tartt, 555 pages

For Keeps:

BOOKS I'LL NEVER
BREAK UP WITH

Once, my husband, Michael, almost ran away with a Subway sandwich artist (this has a point eventually). He was describing the sandwich he wanted, and then suddenly the teenager taking our order was looking up at him with enraptured eyes, the heat coming off of her body so torrid, it popped the top button on her polo shirt.

"That's the *exact* sandwich I eat," she whispered, voice and body atremble. I thought she might clamber over the sneeze guard and into his arms, *Officer and a Gentleman* style.

Though I couldn't be angry, because I know exactly how she felt (eventual point). When, after a long day of discussing James Patterson — cold-cut combo — and Joel Osteen — turkey breast — and requests for self-published e-books about space-alien president conspiracies — those weird Flatizza things — when someone comes in and

302

asks if I've ever read *Weetzie Bat* or requests a biography of Freddie Mercury, I get hot and bothered myself.

One might assume librarians are asked what their favorite books are all the time, but in truth it's pretty rare. Sure, someone may wonder what the last good thing we read was, but they usually don't *really* want to know. What they really want to know is if we're hiding a yet-to-be-released John Green novel in the back.

Thankfully, once in a while, someone does ask what we like to read and really means it. This list is for those patrons. God bless you, and sorry I leaned suggestively over the desk when you mentioned Eugenides.

FICTION

The Virgin Suicides by Jeffrey Eugenides

I mean, you knew this was coming. All. Time. Favorite. I can't help but read this book out loud because it is so beautiful. And wistful. And genius. Plus, it has the heartbeat of my home state of Michigan. And I know I've already talked about it and you've probably (hopefully) already read it, so I'll move on.

Dandelion Wine by Ray Bradbury

The novel I read again every summer for its magic. You can turn to *Dandelion Wine* for any emotion. If you want to feel wistful, grateful, understood, heartbroken, terrified, or tickled, it's all in there. It is the story of a young boy living in a small town in 1930s Illinois realizing how goddamn good it feels to be alive. But it's also a story of every member of his family and the rebellious teens in town, and the old man trying to rig up a machine that produces happiness, and the joy one can get from a push-reel lawnmower, and the local serial killer. It's the best.

Wuthering Heights by Emily Brontë

The book I read every late autumn for its gloom and gothic love. So I can imagine myself standing on a moor with the wind blowing my hair around. Sobbing, obviously.

The Collected Stories by Grace Paley

I dragged a boyfriend with me to hear Paley read so I could get extra credit for a college course and immediately fell in love. With Paley, not the boyfriend. Her stories are both forthright and lyrical. Open this collection to any page, and you will find a line with equal parts skill, wit, and heart, like this line of hers from "Enormous Changes at the Last Minute": "You're not so brand-new. The telephone was brand-new. The airplane was brand-new. *You've* been seen on this earth before." She slays me.

Heartburn by Nora Ephron

Heartburn is probably the funniest novel I have ever read. After I finished it, I listened to Meryl Streep read the audiobook. Because Meryl does everything perfectly, I liked the audio version even more than the book. I never thought extramarital affairs and self-pity could be so hilarious, but they are.

Charlotte's Web by E. B. White

I don't exactly remember, but I think this is the first book I cried over, and I still never kill the spiders in my shower. That's how you know a book has stuck with you.

NONFICTION

Living High: An Unconventional Biography by June Burn

This autobiography of Burn's daring life and travels was written in 1941 and gave me more guidance for living an intentional life than any modern biography. Just be-cause she fucking felt like it, Burn, along with her husband, homesteaded on an island, moved to Alaska and lived in an igloo, and traveled across the United States with a wagon and a donkey AND a kid. I reread this book when I'm pitying myself because my husband didn't make coffee before he left for work and I am forced to parent without caffeine because I don't know (won't learn) how to use the French press.

Random Family: Love, Drugs, Trouble, and Coming of Age in the Bronx by Adrian Nicole LeBlanc

This book goes deep into exactly what it's like to grow up in poverty and without ac-cess to the everyday things most people take for granted. LeBlanc spent ten years re-searching and getting to know a family in the Bronx, following them in and out of homes and prisons and generations. You will know and understand them by the end as well. This should definitely be required reading for public librarians, and then they should make everyone else read it too.

Yes Please by Amy Poehler

An honest (and hilarious) book by a celebrity? Yes. Please. I connected with Poehler over this book so completely that I had to write her a fan letter, even though I knew she would never read it. Amy Poehler, you're my hero.

Living Beautifully with Uncertainty and Change by Pema Chödrön

For hard and unsure times. When I'm feeling Buddhisty or when I just came off of a good shower cry — this is the book I turn to. Chödrön counsels folks to live happily in this unstable world by making peace with that very instability. It's not too self-helpy, it doesn't admonish the reader, and it's written in clear and concise prose. A permanent nightstand fixture.

The Diary of a Young Girl by Anne Frank

I somehow made it through high school, college, and grad school — LIBRARY school — without ever reading this book. And I'm glad that I did because, when I finally picked it up, it was with the appropriate reverence. Anne is the number-one writer I wish I could invite over for dinner, because I would love to watch that bright young woman's smile grow upon learning that she is an internationally celebrated author. Oh, how I wish.

COLLECT THEM ALL!

ANNIE'S ALL-TIME TOP BIOS AND MEMOIRS

Celebrity

American Rose: A Nation Laid Bare: The Life and Times of Gypsy Rose Lee by Karen Abbott

I Am the New Black by Tracy Morgan (Listen on audio, you will be shocked and awed.)

Music

Trouble Man: The Life and Death of Marvin Gaye by Steve Turner

Girls Like Us: Carole King, Joni Mitchell, Carly Simon — and the Journey of a Generation by Sheila Weller

Real Life

The Beautiful Struggle: A Father, Two Sons, and an Unlikely Road to Manhood by Ta-Nehisi Coates
Now I See You by Nicole C. Kear

5 GRAPHIC NOVELS TO READ AFTER *GHOST WORLD*

The Diary of a Teenage Girl by Phoebe Gloeckner
Any Empire by Nate Powell
Joyland by Emily Schultz and Nate Powell
Killing and Dying by Adrian Tomine
Girl Stories by Lauren R. Weinstein

BOOKS ABOUT GIRLS AND ROMANCE THAT DON'T MAKE ME WINCE LIKE *TWILIGHT*

Fiction

Just One Day by Gayle Forman
The Summer Prince by Alaya Dawn Johnson
The Bermudez Triangle by Maureen Johnson
Like No Other by Una LaMarche
Dumplin' by Julie Murphy

310

Nonfiction

Yes Means Yes: Visions of Female Sexual Power and a World Without Rape by Jaclyn Friedman and Jessica Valenti
All About Love by bell hooks
Just Kids by Patti Smith

Epilogue

I thought it was a really good and funny book. The girl seemed smart and funny and I liked her lists. I like ice skating too.
— Annie's Fourth Grade Book Report on *Anastasia Krupnik* by Lois Lowry

Dear Reader,

The above quote is multipurpose. One, because I hope you have similarly bright feelings for *Dear Fahrenheit 451*. Two, it illustrates that, even if you thought this was a shitty book, you have to admit that my writing has improved since fourth grade.

Did *Dear Fahrenheit 451* make you want to reread a book you broke up with long ago? Did you keep notes of all the reading I suggested and now you have a gabazillion books on your list? Do you want to know where you can get a copy of *The One-Hour Orgasm*? You can accomplish all those things and more at — wait for it — your local public library.

The library is genius in its usefulness. It can be a different place for each person who walks in. Your library can help you find a job, go vegan, read up on the new medication you've been prescribed, or learn a new language. Your librarian can listen to your knock-knock jokes or provide a safe space and helpful resources if there is violence in your home. She or he can give you directions to your aunt's house, or tell you if a celebrity is alive or dead, or help you figure out how to give your kid "the talk." All while keeping your ass swimming in books. And movies. And music. And Internet access.

There is no other place where you can go and basically say, "I need help with this area of my life" and someone will respond, "All right, let's figure this out." Maybe your mom, but she's not as good at "going online" as we are. So, please, go discover all the boundless glory awaiting you. And introduce yourself. Otherwise, you're going to get a nickname based on whatever weird face you make or the thing you watched on YouTube that you didn't think anyone noticed.

This is my way of saying: Reader, it's been

fun, but I think it's time we started seeing other books.

All the Best,
Annie

NOTES AND ACKNOWLEDGMENTS

The wonderful Toad-isms quoted on page 67 are from the following stories by Arnold Lobel, in order of appearance: "Spring," "Swimming," "The Hat," "Cookies," "Alone," "Christmas Eve," "Tomorrow," "The Hat."

The plays quoted on pages 75–79 are all selections from volumes 1 and 2 of *Scenes for Student Actors* by Frances Cosgrove. The plays quoted are as follows, in order of appearance: *Small Miracle* by Norman Krasna, *The House of Connelly* by Paul Green, *Nine Til Six* by Aimee Philip Stuart, *Big Lake* by Lynn Riggs, *The Last Mile* by John Wexley, *Berkeley Square* by John L. Balderston, *Something to Live For* by George O'Neil, *The First Legion* by Emmet Lavery, *Lucrece* by Thornton Wilder, *The House of Connelly* by Paul Green, *Accent on Youth* by Samson Raphaelson, *When Ladies Meet* by Rachel Crothers, *Dodsworth* by Sinclair Lewis, and

Rebound by Donald Ogden Stewart.

This book would not, could not be without my agent, Laurie Fox, and a little bit of fate, but mostly Laurie, who liked my writing and took a leap with me and has since provided so much knowledge and encouragement.

Endless thanks also to my editor, Amy Einhorn, who gave me guidance and reminded me to trust my voice, whom I know is the best because every time someone asked who my editor was and I told them, they just about did a spit-take, and to associate editor Caroline Bleeke, who gave me great encouragement and answered each question I had not only promptly and correctly but sweetly. Conveying all these things over email is an art. To the team at Flatiron Books, who made me feel welcome and showed me patience with the occasional gentle push when needed and who worked hard to make this book into A Book.

To my library friends, Heather Booth and Samantha Minnis, who helped me think like a librarian when my brain was mushy. To Kristen Talaga, who partnered with me to make the Fine Print blog a thing, and Brice Bush who supported me and had a secret shelf of ridiculous books for me to cull from.

To my family, who never thought I wouldn't write a book. To Michael and Walden, who left me to type and freeze in the back room and, after I resurfaced, warmed my icy hands in theirs and drew me pictures to make me feel "very much better."

To Daniel Tiger, who allowed me an extra half hour of work without the guilt. To Bruce Willis and the cast of *Moonlighting,* who welcomed me with open arms when I couldn't write another word.

ABOUT THE AUTHOR

Annie Spence has spent the last decade as a librarian at public libraries in the Midwest. She lives in Detroit with her husband and son. *Dear Fahrenheit 451* is her first book.

The employees of Thorndike Press hope you have enjoyed this Large Print book. All our Thorndike, Wheeler, and Kennebec Large Print titles are designed for easy reading, and all our books are made to last. Other Thorndike Press Large Print books are available at your library, through selected bookstores, or directly from us.

For information about titles, please call:
(800) 223-1244

or visit our website at:
gale.com/thorndike

To share your comments, please write:
Publisher
Thorndike Press
10 Water St., Suite 310
Waterville, ME 04901